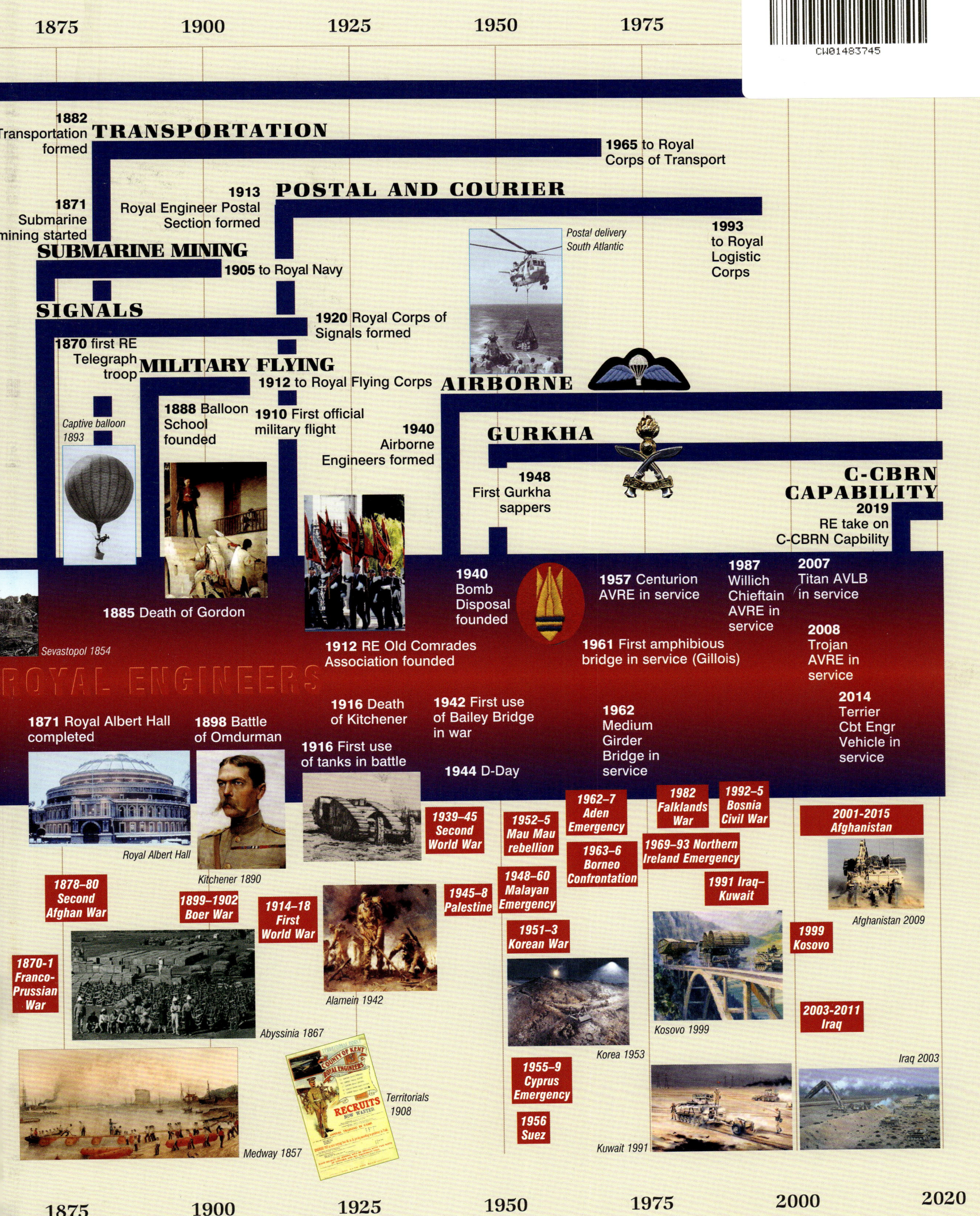

1875   1900   1925   1950   1975

1882 Transportation formed

TRANSPORTATION

1965 to Royal Corps of Transport

1871 Submarine mining started

1913 Royal Engineer Postal Section formed

POSTAL AND COURIER

Postal delivery South Atlantic

1993 to Royal Logistic Corps

SUBMARINE MINING

1905 to Royal Navy

SIGNALS

1920 Royal Corps of Signals formed

1870 first RE Telegraph troop

MILITARY FLYING

1912 to Royal Flying Corps

AIRBORNE

Captive balloon 1893

1888 Balloon School founded

1910 First official military flight

1940 Airborne Engineers formed

GURKHA

1948 First Gurkha sappers

C-CBRN CAPABILITY

2019 RE take on C-CBRN Capbility

Sevastopol 1854

1885 Death of Gordon

1940 Bomb Disposal founded

1957 Centurion AVRE in service

1987 Willich Chieftain AVRE in service

2007 Titan AVLB in service

1912 RE Old Comrades Association founded

1961 First amphibious bridge in service (Gillois)

2008 Trojan AVRE in service

ROYAL ENGINEERS

1871 Royal Albert Hall completed

1898 Battle of Omdurman

1916 Death of Kitchener

1942 First use of Bailey Bridge in war

1962 Medium Girder Bridge in service

2014 Terrier Cbt Engr Vehicle in service

Royal Albert Hall

1916 First use of tanks in battle

1944 D-Day

Kitchener 1890

1939–45 Second World War

1952–5 Mau Mau rebellion

1982 Falklands War

1992–5 Bosnia Civil War

2001-2015 Afghanistan

1878–80 Second Afghan War

1899–1902 Boer War

1914–18 First World War

1945–8 Palestine

1948–60 Malayan Emergency

1962–7 Aden Emergency

1963–6 Borneo Confrontation

1969–93 Northern Ireland Emergency

1991 Iraq–Kuwait

Afghanistan 2009

Alamein 1942

1951–3 Korean War

1999 Kosovo

1870-1 Franco-Prussian War

Abyssinia 1867

Kosovo 1999

2003-2011 Iraq

Korea 1953

Territorials 1908

1955–9 Cyprus Emergency

Iraq 2003

Medway 1857

1956 Suez

Kuwait 1991

1875   1900   1925   1950   1975   2000   2020

 FOLLOW THE SAPPER

FOLLOW THE SAPPER

SECOND EDITION

An Illustrated History of the Corps of Royal Engineers

GERALD NAPIER

with additional chapters by Tom Jackson

and foreword by Major General Mungo Melvin CB OBE

INSTITUTION OF ROYAL ENGINEERS

First published 2005 by
The Institution of Royal Engineers,
Brompton Barracks, Chatham,
Kent ME4 4UG, UK
Second Edition published in 2020 by
the Institution of Royal Engineers.

British Library Publication Data:
a catalogue record for this book is
available from the British Library.

ISBN 978-0-90353-043-9

Text and picture research:
Colonel Gerald Napier
Tom Jackson (Chaps 47 and 48)

Assistant Editor: Captain John Borer.
Museum input: Mrs Rebecca Nash.
Library input: Ms Maggie Lindsay
Roxburgh.
Artwork and captions:
Mrs Hazel Whamond, Ms Beverley
Kelly, Mrs Jacqui Thorndick
Nigel Montagu (Chaps 47–50)

First Edition designed by David
Gibbons

Second Edition designed and edited
by Colonel Nigel Montagu
Assistant Editor: Robin Lloyd Owen

Printed and bound in the UK by
W&G Baird, Count Antrim.

Publisher's Note: Every effort has
been made to ensure the accuracy of
the information presented in this book
but there will inevitably be the occa-
sional conflict of data when comparing
sources. The publisher welcomes
comments and corrections from
readers, all of which will be considered for future editions.

Overleaf: Terence Cuneo's picture
*Breaching the Minefields at El
Alamein*; see page 152.

Right: A sketch by Lance Corporal
Everard at the time of the Boer War;
see page 89.

SAPPERS
RUDYARD KIPLING

When the Waters were dried an' the Earth did appear
 ("It's all one," says the Sapper),
The Lord He created the Engineer,
 Her Majesty's Royal Engineer,
 With the rank and pay of a Sapper!

When the Flood come along for an extra monsoon,
'T was Noah constructed the first pontoon
 To the plans of Her Majesty's, etc.

But after fatigue in the wet an' the sun,
Old Noah got drunk, which he wouldn't ha' done
 If he'd trained with, etc.

When the Tower o' Babel had mixed up men's bat,
Some clever civilian was managing that,
 An' none of, etc.

When the Jews had a fight at the foot of a hill,
Young Joshua ordered the sun to stand still,
 For he was a Captain of Engineers, etc.

When the Children of Israel made bricks without straw,
They were learnin' the regular work of our Corps,
 The work of, etc.

For ever since then, if a war they would wage,
Behold us a-shinin' on history's page –
First page for, etc.

We lay down their sidings an' help 'em entrain,
An' we sweep up their mess through the bloomin'
 campaign
 In the style of, etc.

They send us in front with a fuse an' a mine
To blow up the gates that are rushed by the Line,
 But bent by, etc.

They send us behind with a pick an' a spade,
To dig for the guns of a bullock-brigade
 Which has asked for, etc.

We work under escort in trousers and shirt,
An' the heathen they plug us tail-up in the dirt,
 Annoying, etc.

We blast out the rock an' we shovel the mud,
We make 'em good roads an' – they roll down the khud,
 Reporting, etc.

We make 'em their bridges, their wells, an' their huts,
An' the telegraph-wire the enemy cuts,
 An' it's blamed on, etc.

An' when we return, an' from war we would cease,
They grudge us adornin' the billets of peace,
 Which are kept for, etc.

We build 'em nice barracks – they swear they are bad,
That our Colonels are Methodist, married or mad,
 Insultin' etc.

They haven't no manners nor gratitude too,
For the more that we help 'em, the less will they do,
 But mock at, etc.

Now the Line's but a man with a gun in his hand,
An' Cavalry's only what horses can stand,
 When helped by, etc.

Artillery moves by the leave o' the ground,
But we are the men that do something all round,
 For we are, etc.

I have stated it plain, an' my argument's thus
 ("It's all one," says the Sapper)
There's only one Corps which is perfect – that's us;
 An' they call us Her Majesty's Engineers,
 Her Majesty's Royal Engineers,
With the rank and pay of a Sapper!

CONTENTS

Chief Royal Engineer.

Please convey my warm thanks to the Officers and all Ranks of the Corps of Royal Engineers for their loyal greetings, sent on the occasion of the publication of the Second Edition of "Follow the Sapper" by the Institution of Royal Engineers.

As Patron of the Institution and Colonel-in-Chief of the Corps, I look forward to receiving a copy of the book and much appreciated your thoughtfulness in writing as you did. In return, I send my best wishes to all concerned.

ELIZABETH R.

4th June, 2020.

FOREWORD

It is a great honour to add a Foreword to this new edition of *Follow the Sapper*. Throughout its long and distinguished history, the Corps of Royal Engineers has remained at the forefront of innovation and in the midst of military action. That's why the famous soldiers' cry from the trenches of the Crimean War to 'Follow the Sapper' still resonates today. Wherever and whatever the mission, Royal Engineers have served with ingenuity and valour, blazing an impressive trail. Whether clearing the minefields at El Alamein in 1942, storming ashore on D-Day, 6 June 1944, or opening IED-infested routes in Afghanistan in 2006-14, the same gritty qualities of calm technical competence and resolute personal courage have been on display. Sappers continue to form the vanguard of the British Army in many respects, including their significant role in supporting civil communities and health testing throughout the United Kingdom during the Covid-19 pandemic of 2020.

Gerald Napier's splendid narrative and richly illustrated history of the Corps of Royal Engineers was published to critical acclaim in 2005. Providing a concise and compelling account, it has informed and inspired successive generations of Sappers, whether newly joined or long-serving. Over the last fifteen years, however, much has happened, not least protracted operations in Afghanistan and Iraq, together with a substantial number of humanitarian and infrastructure tasks undertaken worldwide. Hence the time is now ripe to update the remarkable Sapper story.

This volume incorporates two new chapters, 'On a Distant Plain', which focuses on Afghanistan, and 'A Humanitarian Mission', which describes a broad spectrum of military assistance operations at home and abroad. Both have been written by Dr Tom Jackson, the research and editorial assistant to Volume XIII of the Corps History, now under preparation. In addition, the penultimate chapter, 'Esprit de Corps', has been revised by Colonel Nigel Montagu to reflect developments in Corps 'family' affairs over recent years. He has also master-minded and edited this new version of *Follow the Sapper*. On behalf of the Corps, I thank them for their important contributions, as I do Colonel Gerald Napier, whose brilliant original work has stood the test of time so well.

In closing, it would be remiss of me not to mention the two-hundredth anniversary of the founding of the Royal Engineers establishment at Chatham, now the Royal School of Military Engineering, in 2012, and the tercentenary of the Corps, celebrated in 2016. Both occasions reminded Royal Engineers and members of the wider Army and public alike of the Sappers' illustrious history and precious heritage, whether technological, operational or sporting. To help preserve this legacy, and to promote the professional development of the Corps, the Royal Engineers Historical Society has been re-established within the Institution. I urge all readers to take an active interest in the past in order to prepare themselves better for the future, whatever that may bring. Inevitably, Sappers will be somewhere in the lead and others will follow!

Major General (Retd) Mungo Melvin CB OBE FInstRE
Chairman, Royal Engineers Historical Society

PREFACE TO THE SECOND EDITION

The Council of the Institution of Royal Engineers decided to update *Follow the Sapper* to include the dramatic events of the 15 years of operations in Afghanistan. This has been made possible by a generous donation by Colonel David Hindle. We have taken the opportunity to make a small number of corrections to the original text but have maintained the first edition in its entirety and added two chapters. Gerald Napier has again given of his time and expertise to ensure that *Follow the Sapper* continues to be both an attractive and approachable history, and a work of scholarship. Permission to include copyright images in this edition has been sought and we are indebted to those institutions and organisations which have agreed to this without charge.

Colonel N E Montagu FInstRE
Editor

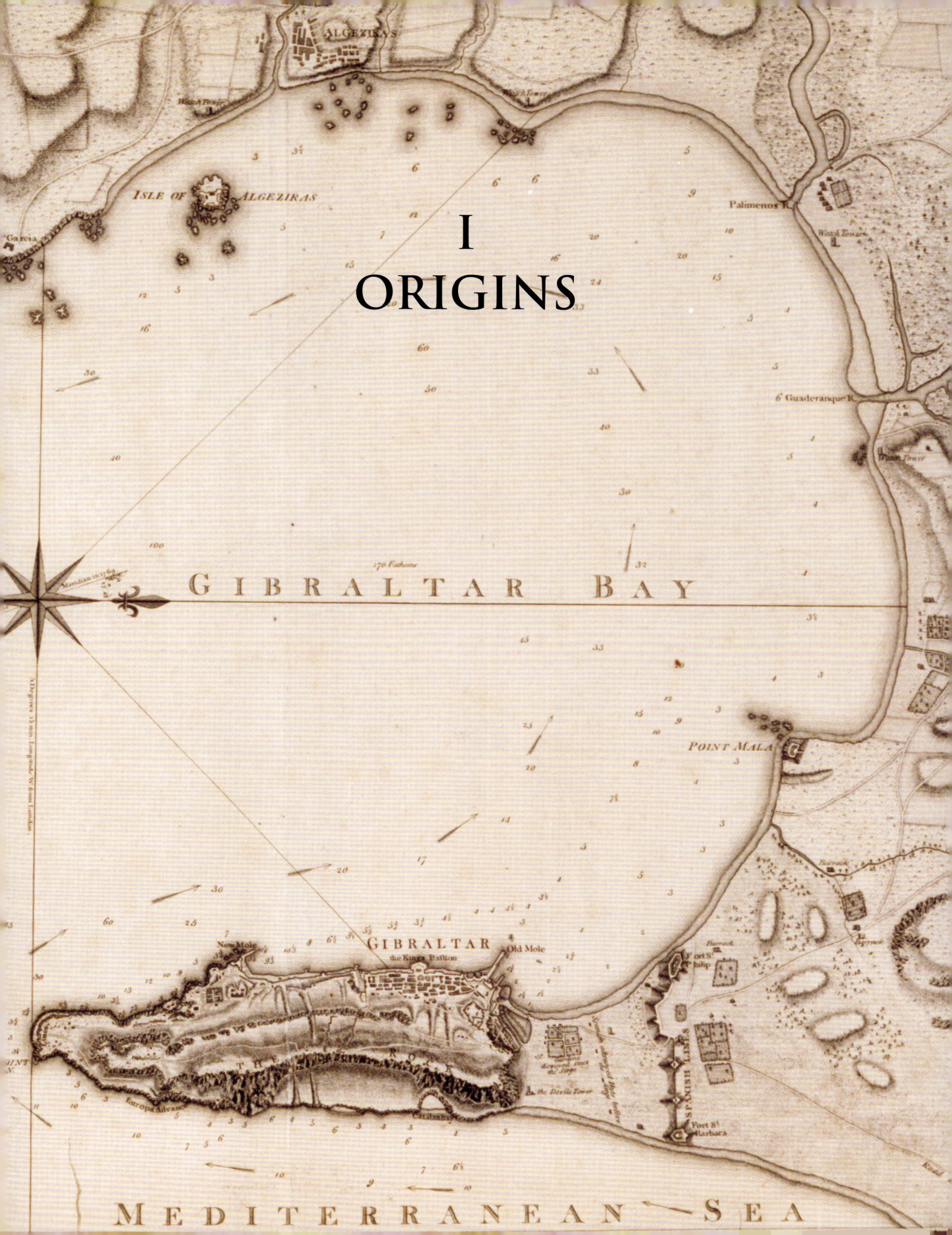

I
ORIGINS

1 | KING'S ENGINEERS

Military engineering secured England for William Duke of Normandy. He arrived at Pevensey on 28 September 1066 complete with a prefabricated fort, which his engineers set up inland at Hastings. We returned the compliment by taking a prefabricated harbour (Mulberry) to Normandy some 900 years later. William knew that to secure his fragile hold over this warlike nation he must immediately build a firm base and thereafter subdue the population by attacking their fortified settlements and establishing a network of his own castles. His Norman successors built their armies around a core of men trained in the arts of war including experts in siegecraft and the construction of forts.

Even before the Normans, military engineering had underlain the success of conquering armies, most notably the Romans. They brought the word *ingenarius* (engineer) with them. The legacy of their skills remains throughout Europe and beyond to this day in the form of fortifications, roads, bridges and artefacts. When they arrived in Britain, they found earlier evidence of the military engineer's art in the Iron Age forts, which took costly battles to subdue. By the time the Romans abandoned Britain they had established

Left: Roman assault bridge. This model is based on the detailed description given by Julius Caesar in his *Commentaries* of an improvised assault bridge thrown across the Rhine in 55 BC. The bridge took ten days to build and was dismantled eighteen days later as the Romans withdrew across the river after a successful punitive attack on the German tribes. (RE Museum on loan from the Royal Artillery Institution)

Above: Maiden Castle in Dorset, with its banks, ditches and tactically sited gateways, exemplifies the Iron-Age forts behind which British tribal chieftains fiercely resisted the advance of the Roman legions.

Above: Motte and Bailey, an artist's impression of the type of structure built by the Normans to control their conquered areas, as it might have looked by the 12th century after development from the original. This comprised a mound (motte), usually itself built on high ground, and one or more palisaded enclosures (baileys). A wood-built 'keep' crowned the motte and this would be approached by means of a drawbridge over a ditch. The bailey would accommodate the main hall, chapel, farm and administrative buildings of the community. In due course the wooden keeps and palisades would be replaced by stone towers and walls

a network of permanent defences and roads in an effort to keep out the hostile Saxons.

Because of his military engineering talents William I's Bishop Gundulf (see illustration) is usually regarded as the 'father of the Corps'

Above; Bishop Gundulf, a monk from the Abbey of Bec in Normandy, came to England in 1070 as Arch-bishop Lanfranc's assistant at Canterbury. His talent for architecture had been spotted by William I and was put to good use in Rochester Cathedral where Gundulf went as bishop in 1077. Almost immediately the king appointed him to supervise the building of the White Tower in 1078. Under William Rufus he also undertook building works on Rochester Castle. Thus he performed the duties of a 'King's Engineer' although apparently never enjoyed that title. Gundulf died in 1108 having served three kings 'and earned the favour of them all'. This statue adorns the west door of Rochester Cathedral.

because the engineer who accompanied William in 1066, Humphrey de Tilleul, defected back to Normandy to sort out his marital problems. After Gundulf's time the title *ingeniator* in the records is applied to distinguish those who were not only skilled builders but also served on the kings' campaigns for siege engine duties. A Waldivius appears in the Domesday Book under this title, but the first about whom much is known is Henry I's engineer, Geoffrey.

When the Plantagenets succeeded the Normans, they continued the practice of establishing their authority through a network of castles and fortified towns throughout their realms on either side of the English Channel. The cost was enormous. Edward I's expenditure on castles in Wales has been compared with a modern government's programme for building a fleet of nuclear submarines. The names of the *ingeniatores* continue to appear

Left: The White Tower, the central keep of the Tower of London, as seen today and **below** as it might have appeared in the 1070s or 1080s while in the early stages of construction under Bishop Gundulf.

Above: Dover Castle, a view of the palace gate entrance into the keep yard looking through the archway of the 13th-century 'Colton's tower' (see also page 246). The castle was developed from earlier earthen ramparts during the late 12th century by Henry II and his engineer, Maurice. A French siege during King John's reign was called off on the king's death. The Parliamentarians took and held the castle during the Civil War. It was extended and modernised during the Napoleonic wars and again played a part in Britain's defence during the Second World War. Then heavy gun emplacements were built for the protection of the harbour, and an underground hospital built together with the headquarters for the Allied commander-in-chief of the 1944 invasion, Admiral Sir Bertram Ramsay.

in the records of these times. Typically Master Elyas was employed both on building works in the 1190s and deployed Richard I's siege engines at Nottingham in 1194.

Warfare at that time being largely centred round the defence of castles, the requirement for experts in siegecraft continued. When Richard's great nephew Edward I invaded Scotland in 1300, the monk, Brother Robert de Ulmo, was in charge of a team of engineers responsible for working the machines. So committed was Edward to engineering that he himself wheeled a token barrow of earth on to the ramparts of Berwick.

At this time Ulmo is referred to as an *attiliator*. The term appears then to have come into use, sometimes synonymous with *ingeniator* but eventually in the 14th century to distinguish the weapons expert (artilleryman) from the engineer. Throughout the Middle Ages until modern times these two specialities, critical to the security of the state, remained firmly under royal control. The arrival of gunpowder on the scene emphasised the necessity of this, leading eventually to the creation of the Board of Ordnance.

Below: Château Gaillard (built 1196–8). Overlooking the Seine at Les Anderlys in Normandy, Richard I's innovatory masterpiece of castle-building integrated a series of barriers into a coherent defence system. It exemplifies the many castles built by the Plantagenets to secure their continental possessions. After his death it was taken by the French king Philip Augustus after a protracted siege during the winter of 1203-4, effectively ending the English hold over their French territories.

Above: Beaumaris Castle, in North Wales, represents the acme of military architecture of its day. Its construction, begun in 1295, was supervised by Master James of St George, a Savoyard in the employ of Edward I, who enjoyed the title 'Master of the King's Works in Wales'. With an inner and outer bailey, it exemplifies the 'concentric' layout with its high inner ring of towers and low outer walls. Modern studies into the interlocking fields of fire from the arrow slots, with far better coverage than in most of the other Welsh castles, have proved the military effectiveness of the design, adding to its manifest demonstration of strength.

SIEGE TECHNIQUES

Machines were the indispensable equipment of an attacking force both to disconcert the defenders and to create a breach through which the fortress could be attacked.

Towers or Belfries

were the main means by which missiles could be hurled over walls. They could also protect the operators of the rams by which breaches could be made.

Mangonels or Petraries (not illustrated)

could hurl stones and other missiles by means of an arm powered by the torsion of a rope skein.

Ballistas or Catapults

fired a stone or a 'bolt' from a giant crossbow drawn mechanically. In the springald the missile was propelled from a frame by two arms powered by vertical twisted ropes or skeins.

Trebuchets

worked on the pivot principle with a long arm carrying° a sling and a counter-weighted short arm. The missile, often a stone but anything unpleasant such as a human or animal corpse, dung or fire ball, would be hurled from the sling. Trebuchets were the siege artillery of the time with ranges around 200 metres and remarkable accuracy.

Mining

under a castle's walls was frequently resorted to. The foundations would be supported on props, which would then be burnt to collapse the walls.

Illustrations from *Grose's Military Antiquities respecting a history of the English Army from the conquest to the present time* (1801) (tower and ram), and Valturius' *De Re Militari* (1483)(trebuchet and springald).

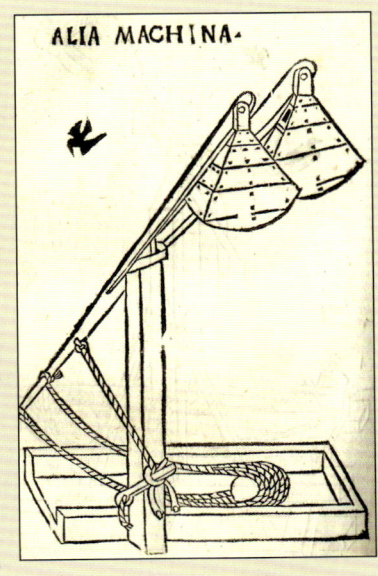

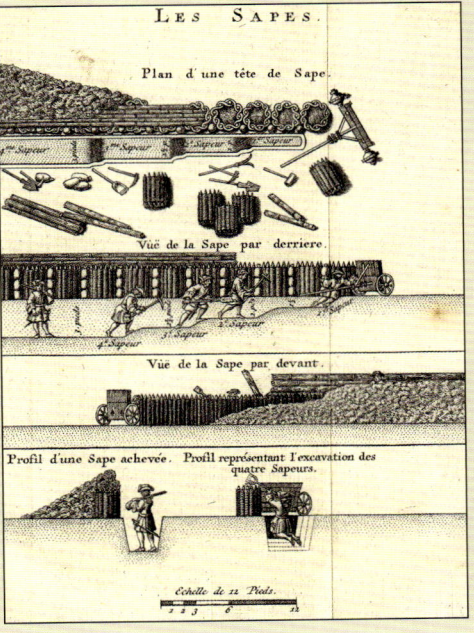

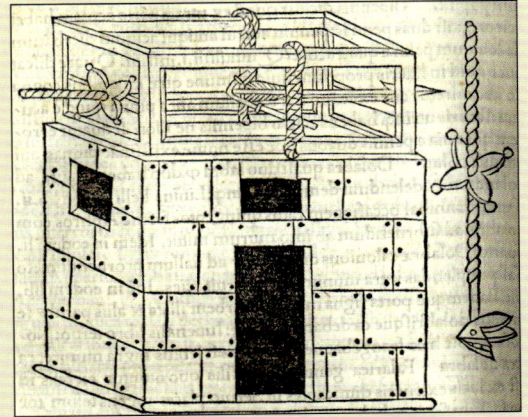

Gunpowder changed everything – but not at once. At the start of the 'Hundred Years War' (1337–1453), in which Edward III launched the long struggle to assert his claim to the French throne and recover control over his continental possessions, he had used cannon at Crécy (1346) and the siege of Calais. Mostly this was against men in the open rather than for breaching fortifications. By the time his grandson Henry V besieged Harfleur in the Agincourt campaign (1415), the emphasis had swung the other way. Sixty-five gunners and cannon, including some of the largest bombards ever seen, with 10,000 'gunstones', significantly contributed to the eventual victory. In command of it all was the engineer, Nicholas Merbury, Master of the King's Works and Ordnance, the first to hold such a title.

The effects on the design of fortifications took longer to mature, at least in England, until the chaotic state of the realm that followed the death of Henry V was resolved when the Tudors came to power.

Henry VIII was obsessed with concerns about invasion from France and ordered numerous

Above: The Board of Ordnance, which until 1855 was responsible for all artillery and engineering matters, was successor to the Office of Ordnance, the department of state that since 1414 had been in charge of the nation's arms and munitions, assuming these responsibilities from the 'Wardrobe' of the Royal Household. The Board was granted arms in 1806, confirmed in 1823 in the following terms: the Arms 'Azure 3 field pieces in pale or on a chief argent 3 cannon balls proper'; the Crest 'Out of a mural crown argent a dexter cubit arm the hand grasping a thunderbolt winged and inflamed proper'; the Supporters 'A Cyclops in the exterior hand of the dexter a hammer and in that of the sinister a pair of forceps resting on the shoulder of each respectively all proper'. (RE Museum, originally displayed in Gibraltar)

SIEGES IN THE 100 YEARS WAR (1337–1453)

Work, work your thoughts, and therein see a siege;
Behold the ordnance on their carriages,
With fatal mouths gaping on girded Harfleur.
... and the nimble gunner
With linstock now the devilish cannon touches,
And down goes all before them ...

Once more unto the breach, dear friends, once more;
Or close the wall up with our English dead.

(Shakespeare, *Henry V*, Act III, Prologue and Scene 1)

Henry V at the siege of Rouen. After the success of Agincourt (1415) Henry extended his control over Normandy in a series of campaigns based on sieges in which Caen, Argentan, Falaise and Cherbourg all fell to him in 1417–18. These were well-prepared with guns and siege machinery such as trebuchets, sows, leather floating bridges and spades, shovels and other essentials. Rouen, the capital, finally fell in 1419. (British Library 1007628.011)

forts to be constructed. His first chief engineer was William Pawne, who had faithfully supported Henry VII during the Wars of the Roses. John Rogers and Sir Richard Lee also served Henry with distinction in a period when military engineering was a growth industry as the king's anxieties about the defences of the south of England grew. Castles were built at Deal, Sandwich, Walmer and Sandown. Blockhouses and batteries were sited at many places around the south coast particularly in Cornwall, Devon and around the Solent. Henry brought in foreign engineers to supplement his own. Lee was with Henry as one of the observers of the *Mary Rose* when she foundered off Spithead during a French attack in 1545, the wreck of which was to

receive the attentions of Royal Engineer divers two and three centuries later (see also pages 110–12).

Offensive actions, such as those against the Scots at Flodden or later in this reign at the siege of Boulogne, were now supported by cannon, maintained in the Tower and deployed in Ordnance 'Trains'. At Boulogne Rogers commanded the Train but his engineers operated separately. At that time engineers were employed by the Office of Works but gradually the Ordnance took on their own. 'Surveyors of Works' looked after fortress work. Mostly this was the maintenance of existing defences, but Sir Richard Lee undertook the building of Upnor Castle for Queen Elizabeth whose reign also saw

Above: Oxford, the great Royalist stronghold of the Civil War, which was finally surrounded and placed under siege by the Parliamentarians in May–June 1646. The Royalist armies had by that time been defeated in the field and it was from Oxford that Charles I set out on his last desperate gamble to put himself at the mercy of the Scots army but which ended with his imprisonment by the Parliamentarians. This painting by Jan de Wyk depicts the defences based on a bastion trace designed by Sir Bernard de Gomme, the Dutch engineer, who had been recruited by Charles and who, after the Restoration, became Chief Engineer of Great Britain from 1661 to 1685 (see page 16). The Parliamentary force under Fairfax is represented in the background behind its line of bastions, located on Headington Hill.

Above: Sir Bernard de Gomme

a flurry of such activity at the time of the Armada. Field engineering, on the other hand, was undertaken on an *ad hoc* basis according to the needs of a particular campaign and was normally managed by officers and men of the army specially allotted to the task, not necessarily professional engineers.

Later, the influence of the Civil War (1642–51) led eventually to a firmer footing for the engineers within the Ordnance service. During that conflict in which sieges, over 200 of them, were the bread and butter of the campaigns, both sides depended much on the engineer talent available. After the Restoration the meagre pre-war establishment of just three engineers was provided for. The deficiency this obviously caused had to be filled from outside the service. A Royal Warrant of 1683 then brought this unsatisfactory arrangement to an end by formally giving the responsibility for military works to the Office of Ordnance. The Board was to be headed by the Master General under whom came the Principal Engineer and the Master Gunner of England. The Principal Engineer had simply a Second and Third Engineer at his disposal, but this embryonic compact band grew rapidly to meet the commitments that multiplied in the following decades as England joined the continental alliances in the wars to contain French expansion under Louis XIV. The War of the Spanish Succession (1702–13), which brought Michael Richards (q.v.) and Holcroft Blood to the fore, was the culmination of this contest.

Above: Sir Bernard de Gomme (1620–85) was Chief Engineer of Great Britain from 1661 to 1685. He was a Dutchman who came to England during the Civil War giving conspicuous service in the royalist cause throughout. Tilbury fort (1670), the modernisaton of the defences at Portsmouth and the Royal Citadel at Plymouth are among his many achievements as Chief Engineer. From 1682 until his death in 1685 he was Surveyor-General of the Ordnance.

FORMATION OF THE CORPS

Almost immediately, the 1715 Jacobite rebellion forced the Ordnance to tidy up its organisation. When he was Principal Engineer, Michael Richards had advocated the formation of the Royal Regiment of Artillery on a separate establishment. This was achieved in 1716, thus taking artillery away from engineer control. The Corps of Engineers was then fixed at a strength of 28, but they were still only quasi-soldiers enduring the hardships and

Below: A 17th-century siege helmet, acquired in 1993, a rare example of a 'pot' helmet of siege weight such as would be worn by gunners and engineers who were particularly at risk in the largely static conditions of siege warfare.

dangers of campaigning in a variety of 'uniform' without the status of military rank unless, as many did, they also held commissions in the Army.

By 1748 the Corps strength had risen to 51 officers (there were still no soldiers). Thirteen years later it was 60. These increases, though woefully inadequate, reflect the burgeoning of overseas garrisons for which provision had to be made and

Above: The battle of Blenheim. The artillery at Blenheim (and Ramillies) was commanded by the engineer Colonel Holcroft Blood who, unlike many of his contemporaries, held army rank throughout his career. Regimental rank was not granted to engineers until 14 May 1757. Blood's career on the engineer establishment had begun in 1688. He was appointed Second Engineer of England in 1696 and in 1702 took command of the Ordnance Train formed for the Low Countries under the Duke of Marlborough. Blood exemplifies the way in which either an engineer or an artilleryman could hold command of an Ordnance Train. Although Marlborough's chief engineer for most of his campaigns was John Armstrong, Blood had won his confidence early on and served him well throughout until his death in 1707.

Above: An Engineer officer of an Ordnance Train, 1702. This figure painted during the 1920s is based on an illustration in MacDonald's *History of the Dress of the Royal Regiment of Artillery*.

Right: The Honourable Brigadier Michael Richards (1673–1721), Chief Engineer of England 1711–14, Surveyor-General of the Ordnance (1714–21), who originally proposed, and eventually oversaw the separation of the Artillery on to their own establishment, away from the control of Engineers, effected by Royal Warrant of 26 May 1716. Richards had served as an Engineer with the Flanders Train in 1692, as an Assistant Quartermaster General in Flanders in 1702 and as Colonel and Chief Engineer of the Train that landed at Alicante.

Above: The Honourable Colonel (later Major General) John Armstrong (1674–1742), seen in this portrait of Marlborough, was the Duke's chief engineer for a number of years and it was his expertise that lay behind the success of many of the Duke's sieges, including his climactic triumph of forcing the 'Lines of Non plus Ultra' and siege of Bouchain. Later Armstrong became Chief Engineer of the Board of Ordnance under the Surveyor General Brigadier Michael Richards, his predecessor as Chief Engineer. Armstrong followed Richards as Surveyor General by which time he was also Quartermaster General of the Forces and Colonel of the Royal Irish Regiment, receiving army pay for these two posts in addition to his Ordnance salaries.

Left: Lieutenant General Sir William Skinner (1699/1700–1780), Chief Engineer of England for 23 years until his death at the age of 81. Skinner grew up in the Corps, having been adopted as a child by his aunt whose second husband was Captain Talbot Edwardes (see also Box 'Theft'). He received his warrant as a 'practitioner engineer' in 1719. He became Chief Engineer at Gibraltar in 1741 having some experience of the Rock from two earlier appearances there. After the Jacobite rebellion of 1745 he was recalled to Britain to undertake works in Scotland culminating in his masterpiece, Fort George (q.v.). He became Chief Engineer of Britain in 1757. Thus he was at the helm during the Seven Years War responsible for providing engineer support to the continental and overseas forces that were put together under William Pitt's dynamic leadership (see pages 26 and 52). Skinner's career began when the Corps had just been established as a distinct body under the Board of Ordnance. He saw its growth into a uniformed military organisation but never became a Royal Engineer.

the many expeditions undertaken. Trains had to be manned in addition to the overseas garrisons. With six to ten engineers on strength, during the War of the Austrian Succession (1740–8), they accompanied expeditions against the Spanish to Cartagena, to Flanders, to North America and to India. Accounts of these conflicts have the stamp of improvisation rather than method, resulting in tales of brilliance and ineptitude in equal measure. The Cartagena expedition, for example, was a disaster largely due to disputes in the higher command, but the lack of experience in siegework among the engineers was the source of some complaint. Highly successful was the capture of Louisbourg in 1745 with the conspicuous help of two engineers detached from the garrison at Annapolis.

Major General John Armstrong (see illustration) had long recognised that the needs of engineering could no longer be

Above: Fort George at Ardersier, Inverness. William Skinner's (q.v.) masterpiece, it was begun in 1748 but not completed until 1769. It has been described as the 'ultimate in promontory forts, it protrudes like a great stone ship into the Moray Firth, with its strongest defences – ditch, ravelin and two great bastions – covering only the landward approach'. It was armed with 81 cannon and contained barrack accommodation for 1,600 soldiers.

left to chance. His reforms, after he assumed responsibility solely for the engineers in 1716, brought a promotion structure and systematic training through experience by attaching cadet and junior engineers to stations such as Minorca and Gibraltar where urgent developments were in progress. Further improvement came with the creation of the Royal Military Academy at Woolwich, 'The Shop', in 1741 and grant of military rank in 1757.

Armstrong's reforms were timely for within a few years even more demands on the Corps were about to be made in America, India and Gibraltar, dealt with separately in this book in the Seven Years War and the War of American Independence (see pages 26–30).

America, India and Gibraltar, which appear separately in this book, loomed large in those great affrays, essentially the Seven Years War and the American War of Independence. The deliberate and lengthy sieges of continental armies had little place in the British experience. Further ahead, in the last decade of the 18th century, the very relevance of the Ordnance organisation was to be put to the test in the crucible of revolutionary warfare (see page 31).

Right: A scissors bridge as conceived by the Comte de Saxe, the Saxon-born soldier who served under Marlborough and later became a Marshal of France. (from Saxe's *Mes Rêveries*, 1757)

Below: Combined Operations 1750-style showing land-ing-craft with ramps, ferrying troops from Gravesend to Tilbury Fort.

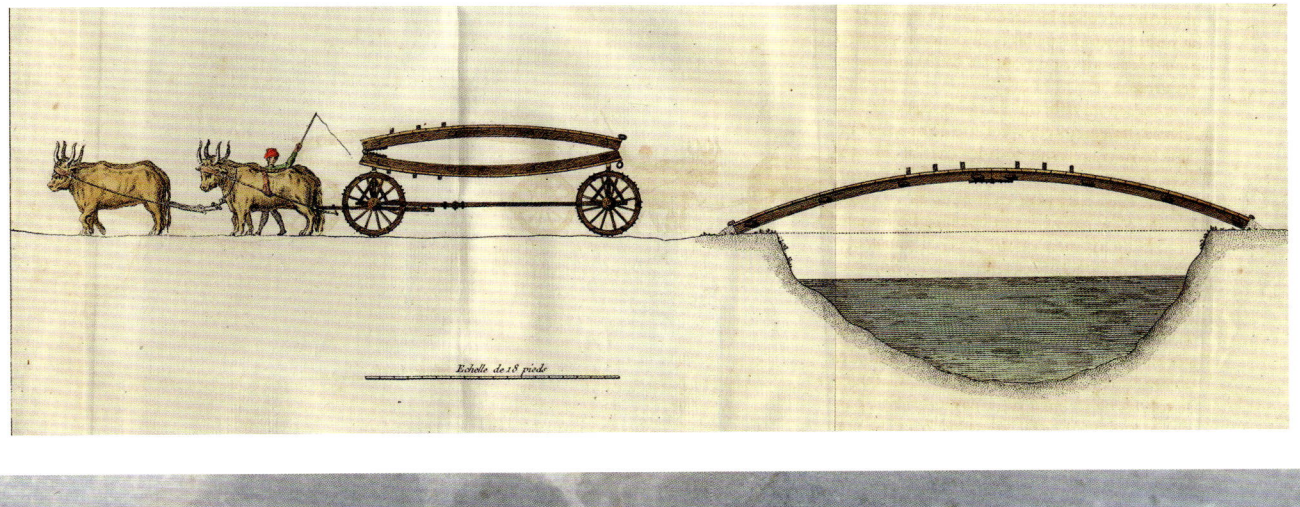

THE SOLDIER-ARTIFICER COMPANY was formed in 1772 on the initiative of Colonel William Green, Chief Engineer Gibraltar, who needed skilled, reliable artisans to work on the urgent programme to upgrade the defences of the Rock against anticipated Spanish attack. They replaced civilians, whose ill-discipline and poor motivation made for slow progress on this crucial work, and were recruited from the regiments serving in the garrison. The success of this new concept of a body of rank and file available to the Royal Engineers prompted the formation of the first six companies of Royal Military Artificers* in 1787 for the express purpose of working on the fortifications in Britain. The Royal Military Artificers evolved into the Royal Sappers and Miners (see page 32), and hence the rank and file soldiers of the Corps originate from the soldier-artificers of Gibraltar. The Royal Military Artificers were also liable for service in overseas garrisons. The soldier artificers (top) are shown in their 1786 red uniform. The Royal Military Artificers (below) originally also wore red but by 1792 were in Board of Ordnance blue. The pictures are from a series of seventeen lithographs produced by Mr George Campion, master of landscape drawing at the Royal Military Academy, for publication in Quartermaster-Sergeant Connolly's *History of the Corps of Royal Sappers and Miners.*

*Originally 'Royal Military Artificers and Labourers'.

Left: Gibraltar, the bay, Rock and town, from a survey carried out by a Royal Engineer, published in 1781.

Below: View of Gibraltar taken from my tent, a watercolour by Lieutenant William Booth, Inspector of Mines during the Great Siege until he was sent home 'insane' due, as he later claimed, to 'lying so long exposed to heat and rains'.

While Holcroft Blood and the Richards brothers were earning renown with Marlborough in continental Europe, around the coast and in the Mediterranean, Britain and her allies were seeking a suitable springboard for operations in Spain itself. This quest eventually brought Admiral Sir George Rooke to Gibraltar at the end of July 1704, with an Anglo-Dutch fleet of fifty ships and a small force, about 2,000-strong, under Prince George of Hesse-Darmstadt. The meagre Spanish garrison was no match for the firepower of the fleet. After a short, stiff, resistance they capitulated and along with some 4,000 inhabitants withdrew across the isthmus while Prince George wasted no time in preparing for the inevitable riposte.

It came a month later when a strong Spanish force with orders to recapture the place arrived and laid siege. Preparations for this expected attack had had to be made without the benefit of an engineer, the first to arrive being Captain Joseph Bennet, on 5 November. He energetically set about improving the defences until the arrival in February of Captain Talbot Edwardes (see page 16), Fourth Engineer of the kingdom, with two more engineers. Edwardes became chief engineer. The garrison clung on, thanks to the support of the Royal Navy, until the besieging force finally withdrew at the end of March 1705. For the remaining eight years of the war Gibraltar thus became a key link in the chain that enabled the fight to be maintained in the Mediterranean. Its acquisition by Britain (along with that of Minorca) was then confirmed by the 1713 Treaty of Utrecht. This did nothing to diminish Spanish determination to recover their lost territory. A further siege was mounted in 1727 on the pretext that Britain had broken her side of the bargain, but it was successfully resisted.

For the next four decades Gibraltar became an icon of British national pride and her imperial aspirations. Concern for its security led to intense

exertions to improve its defences prompted by the arrival of the new Chief Engineer, Lieutenant Colonel William Green, in 1760. Agreement to the works having been obtained, Green then proposed the formation of a body of soldiers trained in the necessary artisan skills, to replace the undisciplined civilian tradesmen who were sent out from England for this work. Thus was born in 1772 the 'Soldier-Artificer Company', whose men were the forerunners of the non-commissioned ranks of the Corps (see page 20). By 1779 Green's prescient assessment of the threat, and that of the two governors he served, Lieutenant Generals Edward Cornwallis (1761–76) and George Augustus Eliott (1776–90), were to prove fully justified.

By 1778 Britain was at full stretch trying to contain her rebellious American colonies. Burgoyne's surrender at Saratoga the previous year encouraged France, and Spain in 1779, to take advantage by declaring war in revenge

Above: The North Face tunnels were started by Sergeant Major Ince of the Soldier-Artificers. In May 1782 the Governor had been musing over the difficulty of countering the fire from the enemy batteries at the north front. He said aloud: 'I will give a thousand dollars to any one who can suggest how I am to get flanking fire upon the enemy's works.' Ince suggested tunnelling to the 'Notch' at the north end of the Rock. As this project progressed, Ince realised that he would have to break through sideways for ventilation. These ventilation tunnels soon developed into gun positions long before the 'Notch' could be reached. Another local brain-child, a 'depression' carriage to allow guns to point steeply downwards, enabled the gunners to capitalise on Ince's work.

Above: Lieutenant General Augustus Eliott (Governor 1776–90) and Colonel William Green (Chief Engineer 1760–83) shown during the Great Siege. Eliott's army career had begun as an engineer for which he held a warrant from 1741 to 1756. He and Green had served together in the Flanders Train in 1745 and were both at the battle of Fontenoy. Eliott had previously served in the Prussian and Austrian armies and acquired a commission in the Horse Grenadier Guards. He was a tough, frugal man, sparing with sleep and inured to the hardships of war. Green was likewise an experienced soldier. He served in several European actions and at Louisbourg and Quebec in the Seven Years War. He was the architect of the King's Bastion as part of the substantial improvements to the defences of Gibraltar before the Great Siege and the instigator of the Soldier-Artificer Company. Not surprisingly, relations between the two men were often strained during the siege. Green would not have appreciated Eliott's professional interference in the Chief Engineer's business. It is said that the original of this picture shows Green saluting the Governor but that Green later asked that the salute be painted out. However, the two men achieved a highly effective working relationship in the common cause of vigorously repelling all efforts by Spain to gain a foothold on the Rock. Both were duly honoured for their work, Eliott as Baron Heathfield of Gibraltar and Green with promotion to brigadier-general and a baronetcy. Mrs Green was present throughout most of the siege and kept an interesting diary in which she recorded the domestic tribulations of the families of the garrison and local population, including the effects of the severe outbreak of smallpox. She fell ill, probably from a chill caught from time spent sheltering in her 'bombproof', and had to be evacuated to England in 1782 but died shortly after.

Above: The Picton Medal is an example of one privately struck to commemorate an action before the days of campaign medals. Lieutenant Colonel William Picton commanded the 12th Regiment during the Great Siege and may have caused this silver medal to be made and issued at his own expense, to commemorate the successful defence of 1782. It depicts a view of Gibraltar and carries a long inscription ending: '… The Garrison under the Auspices of George III Triumphed over the Combined Powers of France and Spain.' The medal was presented to Sir William Green and remained in his family until purchased by the RE Museum.

for their losses in the Seven Years War. For Spain this meant recovering Gibraltar. Two men stood in their way: Eliott, the future Lord Heathfield, and Green.

For four years Eliott inspired the garrison of 5,400 and civil population of 3,200 into a stalwart and pro-active defence. For much of the time the besieging force contented themselves with maintaining the blockade, expanding the siege works on the isthmus and sustaining bombardment of the defences and the town itself. At times this regular onslaught amounted to some 1,500 rounds per day. Thus, presumably, they hoped so to debilitate the garrison as to force surrender or reduce it to a state of incapability to resist an assault. The defenders' morale rose and fell with the rate at which supplies were able to penetrate the blockade. One major delivery was achieved by the Royal Navy in April 1781 at the expense of reducing their blockade of Brest and so allowing the escape of the fleet under Admiral de Grasse that led to his victory off Chesapeake Bay and the surrender of Cornwallis at Yorktown.

A feature of the defence was the ingenuity with which each new problem was met. Green's team comprised ten Royal Engineers, at least three other officers from the regiments of the garrison as assistant engineers and 114 soldier-artificers. They worked with the artillery to devise methods of delivering red-hot shot on to their targets. The artillery themselves perfected an accurate method of fuzing to deliver airburst and light shells. Famously, the soldier-artificers created the tunnels in the north face of the Rock that allowed the new 'depression' carriages to bring flanking fire on to the enemy siege lines. A brilliant sortie by the garrison in September 1781 in which two Royal Engineers and fifty-two men of the soldier-artificers participated exemplifies the aggressive attitude of the defence. It resulted in severe disruption to the enemy lines. But the ultimate test came on 13 September 1782 when the great Spanish attack initiated by ten floating batteries was brought to a swift and decisive end (see page 24). The doughty resistance had proved Gibraltar's ability to withstand a siege and secured the Rock indefinitely into the future for Britain.

Thus Gibraltar took its place as the essential base for operations and naval support that

The King's Bastion was a major feature of the new defences built along the western shoreline of Gibraltar by the Company of Soldier Artificers during the 1770s. Designed by the Chief Engineer, William Green, construction began in 1773 and lasted three years. Eight hundred men lived in the bastion. Twelve 32-pounder guns and four 10-inch howitzers were mounted on the seaward faces. Ten smaller guns and howitzers pointed from the flanks. This model was made by Sergeant Shirres and Corporal Brand of the Soldier-Artificers while young men waiting to join the unit. (RE Museum on loan from the Royal Artillery Institution)

Left: Floating batteries formed the centrepiece of the Franco-Spanish grand attack on 13 September 1782. Ten specially modified ships of the line were each given an extra hull, a sloping roof for deflecting cannon balls and a built-in pumped-water fire-prevention system, and armed with up to twenty guns. They were to sail into position under jury rig to subdue the defenders' batteries, particularly on the King's Bastion, and create breaches for assaulting troops. The attack began at 10.00 a.m. The garrison responded with red-hot cannon balls. At first their superhuman efforts manning the furnaces in the hot weather appeared to be having no effect beyond a little smoke spotted from one of the vessels in the early afternoon. In fact fires breaking out deep in the hulls began to take effect not long after midnight, growing in ferocity. In the ensuing chaotic hours some 2,000 Spanish sailors were lost struggling to escape the flames, despite the heroic efforts of the Royal Navy to rescue as many as possible. This disaster for the Spanish effectively ended the Great Siege.

proved its worth during the Peninsular War and throughout the 19th-century expansion of Britain's empire. The engineer presence was maintained both to support the garrison and to continue the development of the defences in line with advances in technology: the sea-wall line, the 'retired' batteries higher up the slopes of the Rock and the galleries that continued to be burrowed into the north face to cover the landward approaches. The strength of the engineers remained steady throughout the last part of the century at four fortress companies and four or five officers on works, all under the command of a colonel Chief Engineer.

The turn of the 18th century saw consolidation of the strength of this now formidable stronghold. The dockyard was modernised with three dry-docks and the facilities of the naval base improved. However, apart from its value as a support base, for example, providing hospital facilities for the wounded from Gallipoli, Gibraltar played little part in the First World War and was never itself seriously threatened.

Nor was it in fact really endangered in the Second World War although the possibility was more imminent, considering that Italy was part of the enemy Axis, and Spain neutral but tending

to cooperate with Hitler. However, Gibraltar as a base for Admiral Somerville's Force H was very much in the front line. The Rock became a military fortress, all non-effective civilians being evacuated and the military presence built up to nearly 18,000 including a plethora of specialist engineer units organised into two brigades: ten companies (fortress, artisan works, quarrying, road construction, excavator and army troops) and the all-important 3rd Tunnelling Group (four tunnelling companies, a general construction company and a detachment of Canadian specialists in diamond drilling).

The tunnelling for accommodation, a hospital, workshops, stores and all the paraphernalia associated with a large force was one of the Corps' great achievements of the war. Nissen and other huts, specially treated to resist the effects of high humidity, were installed underground, along with the services such as electricity, water supply and ventilation. By the end of the war the total amount excavated was calculated to be the equivalent of a ten-foot-diameter tunnel from London to Liverpool.

The other great project was the airstrip. A 900-yard runway across the isthmus built quickly after the outbreak of war soon proved inadequate. It was eventually extended to 1,800 yards by building into Algeçiras Bay. The Gibraltar runway reduced dependence on Malta as a staging post and was ready for limited operations in support of the 1942 campaign in north-west Africa although it did not reach its full length until 1945.

The garrison was rapidly reduced in the post-war years, the sapper element to a fortress regiment by 1960 and reducing to a squadron by 1961. Although the wartime tunnels were gradually abandoned, tunnelling continued, mainly to improve traffic flow round the Rock, by means of a tunnelling troop that remained part of the squadron. Soon this work, too, ceased to be a Corps responsibility. A small residual presence remained until in 1994 the last sapper unit left Gibraltar, leaving behind a statue of a soldier-artificer of 1772 in commemoration of the role that Gibraltar had played in the evolution of the Corps.

Above: The Great North Road (above), one of the communication tunnels built during the Second World War. Chambers, normally from 28ft to 40ft span and 150ft to 200ft long were excavated off the tunnels for housing the multiplicity of facilities such as headquarters, workshops, hospitals and living accommodation for the garrison.

Left: Sappers working on the entrance to 'Arow' Street tunnel (on the south-eastern side of the Rock). Arow Street was named after the CRE, Lieutenant Colonel A. R. O. Williams. (IWM GM43)

Below: The Gibraltar airfield extension was decided upon in December 1941. The construction units arrived early in 1942 and the extension was ready for use in support of the Torch operation in November of that year although it did not reach its full 1,800-yard length until 1945. (IWM GM1086)

4 | AMERICAN ADVENTURES

Three wars in the short space of sixty years established Britain's relationship with North America for the two centuries ahead. The 'Seven Years War' (1754–63*), known in America as the 'French and Indian War', not only established Britain's dominance over that area but, elsewhere round the globe, founded much of the British empire. The War of American Independence (1775–83), however, clipped Britain's wings, leaving her only Canada over which Americans felt they should have some rights (see page 29). The War of 1812, however, resolved the position of Canada.

THE SEVEN YEARS WAR IN AMERICA

British strategy in America in 1754 aimed to mobilise the colonists to support an expeditionary force to block French encroachment south from the Great Lakes. In a classic stumbling start, two-thirds of a 1,500-strong army under General Braddock was destroyed in ambush on the Monongahela River. Four of the five engineers were wounded, including their commander, Engineer in Ordinary Patrick Mackellar, deputising for his chief Sub-Director James Montresor who was ill.[†] Mackellar had supervised the creation of the route through tough country over mountains and across marshes and rivers. Matters improved as William Pitt gained control over the home government and appointed a new generation of commanders who gradually brought a succession of victories round the world. In America, while a few key episodes remain the most memorable features, the war was won as much by relatively minor infantry operations to gain control of the frontier, conducted over long distances through rugged country in all weathers, often harassed by hostile Indians. Each column would normally have its own engineer whose tasks would include blazing the trail, creating improvised

*Named by the French 'La Guerre de Sept Ans', it actually lasted nine years.

†These ranks became major and captain respectively when military rank was granted in 1757.

Below: The layout of the new town of Georgetown (now Es Castell), Minorca, drawn up by Colonel Patrick Mackellar who became Chief Engineer after Minorca had been returned to British hands following the Seven Years War. This post was some recognition of his outstanding service in America, at Quebec and in the West Indies. **Inset:** The inscription on the plaque illustrated reads 'Es Castell dedicates this street to Patrick Mackellar Military Engineer (Argyleshire 1717– Es Castell 1778) originator of the urban plan of our town. April 2002'.

fortifications and supervising siegeworks against enemy forts.

By 1758 Pitt's resolute strategy was bearing fruit as an overland column made its way north up the Hudson valley, aimed at Montreal. At the same time the French fortress of Louisbourg on Cape Breton Island, Nova Scotia, guarding the entrance to the St Lawrence, was taken by a force under Major General Jeffrey Amherst after an eleven-week siege conducted by ten engineers under Colonel John Bastide. The intemperate General James Wolfe, who commanded a brigade in the attack, criticised the work of the engineers:

> The parapets [of Louisbourg] in general are too thin, and the banquettes every where too narrow … It is impossible to conceive how poorly the engineering business was carried on here. This place could not possibly have held out ten days had it been attacked with common sense.

Bastide was a somewhat elderly plodder but Wolfe's censure typifies the tension between the impetuous commander and methodical engineer. In this case the garrison surrendered without the necessity of a bloody assault and the overall British casualties were only 523 killed and wounded from a force of some 12,000 men.

Major Patrick Mackellar replaced Bastide and was Wolfe's Chief Engineer for the attack on Quebec the following year (1759). Two of his captains, William Green and Hugh Debbieg, were to achieve distinction in their later careers.

Above: Captain John Montresor (left, as captain; right, as colonel) was a skilled engineer who took an active role in the American War. He was the son of James Gabriel Montresor who also served in America. In the Seven Years War father and son were in the force under General Abercromby that failed to take Fort Ticonderoga in 1758. Later young John distinguished himself by volunteering to take the despatches reporting the fall of Quebec to Lord Amherst, a 26-day journey in which he suffered cold, hunger and physical danger from hostile Indians. He remained in America and, on the outbreak of the War of Independence, was present at the battles of Bunker Hill and New York. Later he moved to Philadelphia where he planned the construction of Fort Mifflin on Mud Island, designed to control navigation of the Delaware River. But the events of the war forced the British to evacuate the city. So in 1777, six years after his original work, he found himself back at Philadelphia in charge of the siege of the fort that he had planned but had been built by the 'rebels' in the meantime. He resigned from the Corps in 1779 and lost his life at sea in 1805 while on passage from India to Malaya.

Left: A Plan of the City of New York and Its Environs, surveyed by Captain John Montresor, showing the city with fields and hills to the north-east. Inset top left is a map of the area.

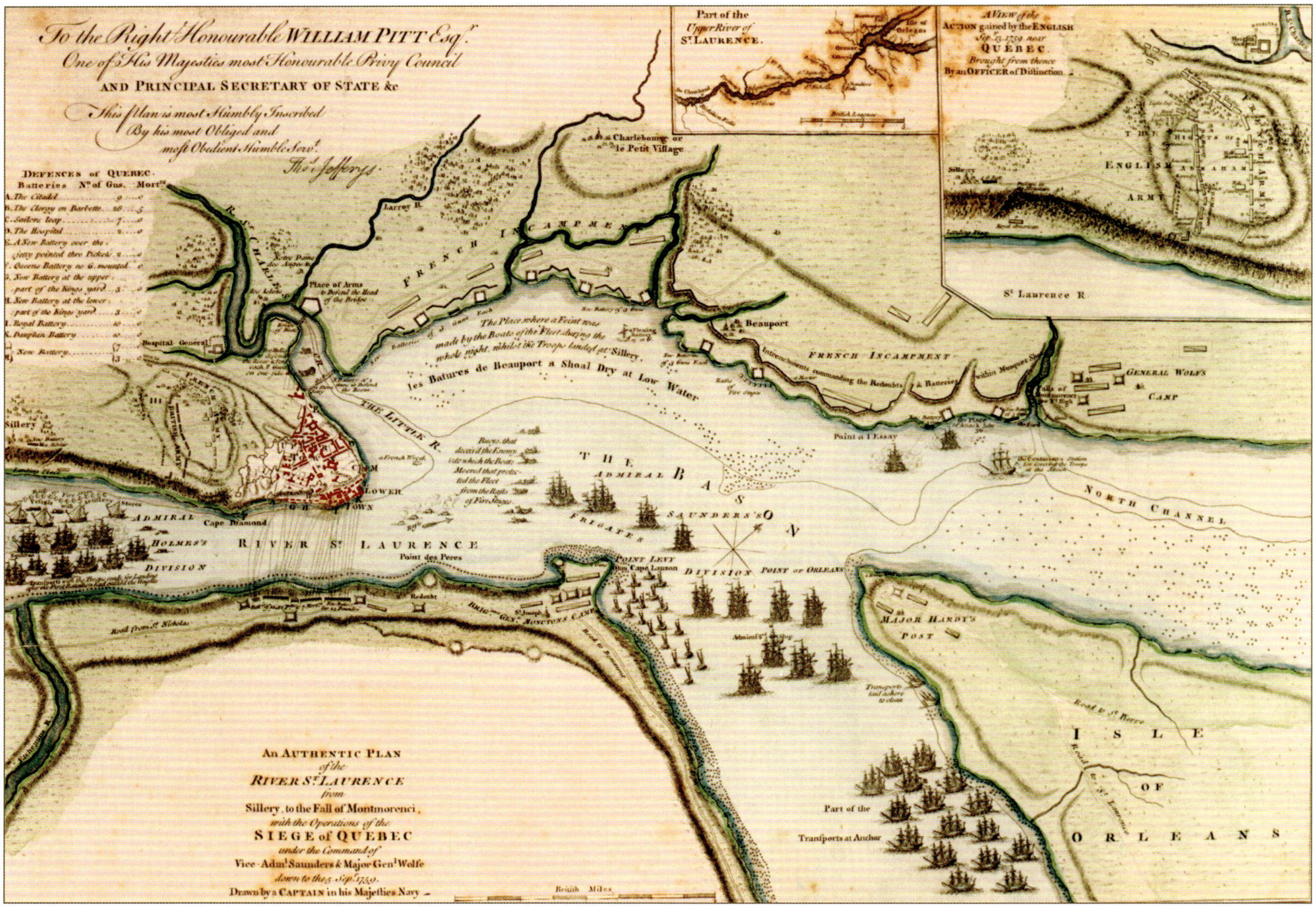

Left and below: Fort Ticonderoga was built by the French (who called it Fort Carillon) to cover the north–south route from Montreal through Lakes Champlain and George. It typifies the many strategically sited frontier forts in the disputed border area. In 1758 an army under General Abercromby failed to capture it from the small French garrison under General Montcalm. The 42nd Highlanders suffered 203 killed and 296 wounded attempting to break through an abattis. The next year the French evacuated the fort without a fight when confronted by an overpowering force under General Amherst. In 1775, early in the War of American Independence, Ticonderoga was captured by the Americans in a dashing coup to acquire its guns. It then became a base for their failed expedition to Montreal and Quebec to rally the Canadians to their side. Two years later it was taken back by the army under General 'Gentleman Johnny' Burgoyne, en route to its fate at Saratoga. This watercolour was painted by one of Burgoyne's engineers, Lieutenant Henry Rudyerd. The bridge across the lake had been removed to allow vessels to pass and the remains of the piers are visible.

Wolfe's efforts to take the well-fortified city lasted from 26 June to 13 September. During that time the main engineer business was the preparation of batteries. From his despatches it is clear that Wolfe was happy to consult Mackellar over tactics and appreciated the work of his engineers. Mackellar himself kept a journal from which this extract describes the climax as Wolfe lay dying on the Heights of Abraham:

> When they were within 100 yards of us our line moved up regularly with a steady fire, and when within 20 or 30 yards of closing, gave a general one, upon which the enemy's line turned their backs from right to left in the same instant. They were by ten o'clock pursued within musket shot of their walls, and scarce looked behind till they had got within them.

Later Mackellar commanded the substantial force of engineers which took part in the successful sieges of Martinique and Havana that rounded off a profitable series of campaigns in the West Indies to relieve both the French and Spanish of their possessions.

By the end of the Seven Years War the Corps establishment had increased to 62 officers. While the status of army rank had been achieved, engineer grades were also used in official lists, for example Director and Lieutenant Colonel, Sub-Director and Major, Engineer in Ordinary and Captain, Engineer Extraordinary and Captain-Lieutenant, Sub-Engineer and Lieutenant and Practitioner and Ensign. These professional titles were not abandoned until 1782.

THE WAR OF AMERICAN INDEPENDENCE

The ten-year abortive struggle to retain the American colonies involved no major engineering projects. In the wide-ranging campaigns the engineers marched with the army to supervise the field engineering that each situation called for. After the pyrrhic victory at Bunker Hill (1775), at which the engineers Captain John Montresor and Lieutenant Thomas Page were present, things looked more promising as General Howe ejected General George Washington's forces from Long Island, Brooklyn and Manhattan Island. But he was unable to follow up these successes and Washington escaped. John Montresor was also

present for this campaign.

Fatal lack of coordination between the British commanders led to the surrender of General 'Gentleman Johnny' Burgoyne's army at Saratoga in 1777.[*] British hopes of success then turned to action in the southern states where they had established a base at Savannah and where there was substantial support among the colonists. The French had now entered the war and, with a powerful fleet, supported the Americans in besieging Savannah. They gave up after six weeks, much of the credit for the defence going to Captain James Moncrieff:[†]

> The zeal and talents of Captain Moncrieff, the Chief Engineer, and the unremitting exertions and labour of the officers and soldiers, assisted by the negro population, completed a line of redoubts … the French officers declared that the English Engineer made his batteries spring up like mushrooms.

Moncrieff also distinguished himself in the successful British siege of Charleston the following spring (1780). Despite this severe setback for the American cause, the writing was on the wall. The French with their powerful fleet had tipped the military balance. Washington's resolute leadership over ten years had inevitably turned the mood of many 'loyal' colonists and the inevitable showdown duly took place at Yorktown with the surrender of Cornwallis' army on 19 October 1781.

THE WAR OF 1812

Just as the French had acted to exploit Britain's overstretched military capability in 1777, the Americans took advantage of the situation in 1812 to try to snatch the Canadian provinces and so secure their northern frontier. The war on land was generally evenly balanced. Engineers contributed to two notorious episodes. The August 1814 raid on Washington in which many public buildings were burned was a military success but a political disaster. Captain Blanshard,

---

[*] Burgoyne was the father of the first field marshal of the Corps, John Fox Burgoyne, by his mistress the opera singer Susan Caulfield.

[†] Moncrieff was one of the brighter stars in the engineer firmament at the start of the Revolutionary Wars. Unfortunately he was killed in action in Flanders in 1793, quite possibly the victim of mistaken identity due to wearing his blue jacket without the white armband that it had become the practice for officers of the Ordnance to adopt to avoid confusion with the French.

later to become known as the designer of a successful pontoon (see page 43) led the engineer contingent from the 2/4th Company of the Royal Sappers and Miners. At least they were able to enjoy the feast that President Madison had arranged for his anticipated celebration of a British setback.

More consequential was the offensive against New Orleans, an attempt to grab some British advantage out of the war. It involved six thousand men under General Edward Pakenham with Colonel John Fox Burgoyne, eleven other Royal Engineers and two companies of sappers (2/4th and 7/1st). A frontal assault across marshy exposed ground and against excellently prepared earthworks failed at great cost, including Pakenham's death. The supporting left flank attack devised by Burgoyne and involving some enterprising river-crossing work by the

BURGOYNE'S SECRET

After Pakenham's death at New Orleans (see text), the senior officers present met to decide on the next step. The only available meeting place was a small building in the middle of which was a table on which lay the body of a member of the staff, covered with a sheet. They decided to abandon the assault despite the promising development of the left flanking attack. Burgoyne was given the task of withdrawing the troops, which he successfully accomplished. This gave rise to the belief, in later discussion of the whole episode, that Burgoyne had been the influential voice in the decision to withdraw. In fact the reverse was the case. He had urged the renewal of the attack. However, he never said anything to counter the slur. His true contribution to the discussion only came to light because the body on the table was that of Major Stovin who had been declared dead after a hurried examination but was in fact only wounded, having been shot in the throat. He had heard the whole discussion and years later the then Major General Sir Frederic Stovin told the story in confidence to the sapper Major General Sir John Cowell.

Left: Colonel John Fox Burgoyne from a portrait painted by Thomas Heaphy towards the end of the Peninsular War.

sappers, was called off after the failure of the main assault. Some honour was then restored by the capture of Fort Bowyer, Mobile, but news that a treaty to end the war had been signed at Ghent prevented any advantage being taken of this, the last offensive British action on American soil.

Left: The Citadel at Halifax. The War of 1812 drew attention to the vulnerability of Upper Canada to attack from the United States. The Citadel, completed 1856, is the fourth in the series of forts built for the protection of this key strategic naval station. Its purpose in its present form was for defence from attack by the United States.

5 | COMING OF AGE

The Corps was transformed by the revolution that swept Europe and the world at the end of the 18th century. In just a quarter of a century, a loose association of individuals grew into a mature, professionally directed and purposeful body.

Royal status had been granted to the Corps of Engineers by a warrant of 25 April 1787. Six months later the Corps of Royal Military Artificers was created after the model of the soldier-artificer companies in Gibraltar, with the prime purpose of maintaining the fortifications in the garrisons of Portsmouth and Plymouth with companies based there, and at Woolwich, Chatham, Gosport and the Channel Islands.

This frail structure amounting to seventy-one officers and 600 soldiers now had to cope with the whirlwind that followed the French Revolution. Fortification was its stock in trade. Of the arts of field engineering these bodies were largely ignorant. Nor was their higher command well equipped for such an undertaking. Both corps lay within the realm of the Master General of the Ordnance who allocated them, as directed by the Secretary of State, to the Commander-in-Chief in the field. Moreover, promotion was by seniority and the higher ranks of the Corps lacked the nimble minds and eager energies that the situation demanded. This unpreparedness for taking on revolutionary France applied also to the rest of the army, though for different reasons. The weaknesses were about to be exposed.

THE REVOLUTIONARY WARS (1793–1802)

After a brief sortie into continental Europe under the Duke of York in 1793, Britain fought France largely at sea and overseas. The West Indies became a key theatre of the war in 1794 where the young Richard Fletcher (see page 34), cut his operational teeth. Campaigns followed in Corsica and Egypt, each giving a handful of Royal Engineers experience that was to pay off

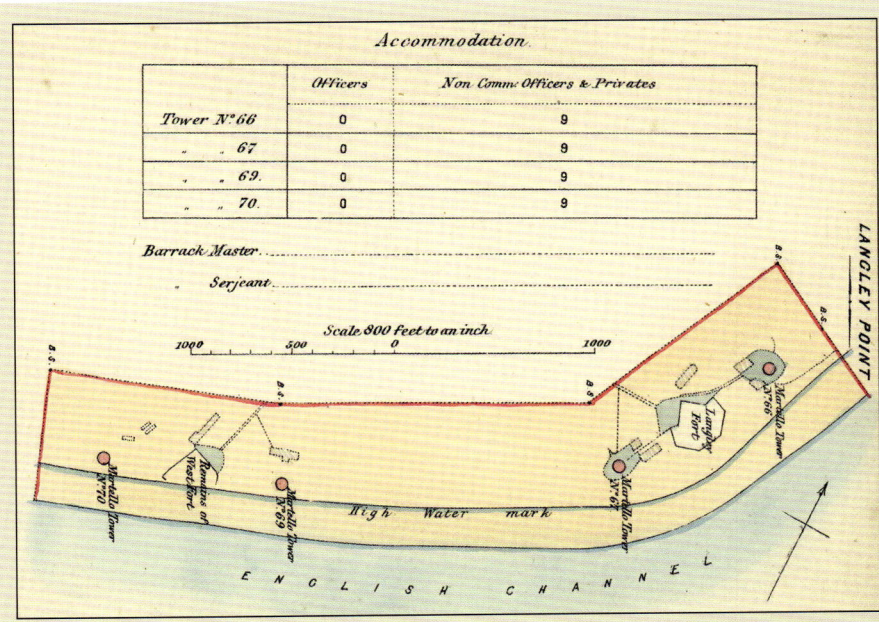

MARTELLO TOWERS were based on the design of a fort at Mortella Point in Corsica, which in 1794 had held out for three days against a British infantry assault supported by naval bombardment. In the threat of Napoleonic invasion, 164 were built in Britain, mainly on the south-east coast but also in Ireland, Jersey and the Orkneys. Their design was initiated and construc- tion supervised by Royal Engineers. The Tower illustrated is No. 73, colloquially known as 'The Wish Tower', which still stands at the west end of Eastbourne promenade. The lithographed map, illustrating the spacing of these towers on the south coast, is taken from one of a set of pocket books of 1863 listing all War Office installations.

in the years to come. The threat of imminent invasion at this time gave urgency to improvments in home defence so that many engineers and the Royal Military Artificers were involved in improving fortifications.

The Peace of Amiens (1801–3) brought a brief relief from the fighting, but not before one last British fling in continental Europe, the misconceived Helder expedition. It failed but was notable for the creation of the third military engineering

THE CORPS OF ROYAL MILITARY ARTIFICERS AND THE CORPS OF ROYAL SAPPERS AND MINERS

The Corps of Royal Military Artificers (top, working dress) was formed in 1787 on the model of the Gibraltar Soldier-Artificer Company, to reduce the cost of upgrading the defences of Great Britain for which the 3rd Duke of Richmond, Master-General of the Ordnance from 1783, was the inspiration. The idea of a corps of artificers drawn from the army had been proposed by Lieutenant Colonel Hugh Debbieg in 1779 but was rejected by the then Commander-in-Chief, Lord Amherst, who considered that '… the general business of the pioneers must

be done by the able-bodied men amongst the peasants of the country'. The Duke's proposal would be for a new corps under the Ordnance, so saving the cost of civilian artisans. It was strongly opposed in Parliament. It was argued that the concept of placing civilian tradesmen under military discipline was a repugnant infringement of rights. Nor was it believed that well-paid artisans would volunteer for the same work at significantly less pay. Nevertheless, the measure was negotiated through Parliament by William Pitt, and six companies were raised for

Woolwich, Chatham, Portsmouth, Gosport, Plymouth and Guernsey. The new corps had its own officers, known as sub-lieutenants, responsible for discipline and administration but for work they were commanded by Royal Engineers. Overall command was vested in the senior Royal Engineer in the garrison and the corps as a whole had a Colonel Commandant, the first of whom was Major General Sir William Green of Gibraltar fame.

The scheme was generally successful but by the outbreak of war in 1793 some of the soldiers had become reluctant to exchange their comfortable billets with their families together with some profitable 'moonlighting', for the rigours of war. Nevertheless, many of those who did so earned distinction and honour, including at the 1793 siege of Valenciennes, despite the lack of training in field engineering skills.

This deficiency became more apparent as the war went on until experience at the sieges of the Peninsular War finally persuaded the senior command that a permanent body of men trained as sappers and miners had to replace the system of reliance on infantry of the line for field engineering.

The Corps of Royal Sappers and Miners (below) was created in 1813 to meet this need. No extra manpower was provided. The Royal Military Artificers were simply renamed. However, by that time they had an establishment of nearly 3,000 compared with the original 600. Moreover, the four companies (about 500 men) in Portugal and Spain had been trained hard in field skills and learned their art in battle. In England the Plymouth company was converted to its new duties by being marched to Chatham under its commander, Major Charles Pasley (see page 42), who had now been authorised to set up the necessary training course. They were ready in time to take part in the siege of San Sebastián in September 1813 where they arrived in their new scarlet uniforms and became known as 'Pasley's Cadets'. The change in colour of the uniform from the 'Ordnance' blue to the scarlet more general in the army was also proposed about this time for officers although not immediately adopted. There had been several incidents of sappers in blue being fired upon by their own side, mistaken for French, including Colonel James Moncrieff, Chief Engineer and Quartermaster General of the army, at Dunkirk in 1793.

arm to appear in this story. The brainchild of the Duke of York, the Royal Staff Corps was designed to provide the army with a field engineering capability directly under the Commander-in-Chief and independent of the Ordnance. It attracted much talent and was to perform some brilliant feats until its responsibilities were absorbed by the Royal Engineers in the 1830s.

THE NAPOLEONIC WARS (1803–1815)

These adventures had toned up Britain's military muscle for the great test ahead. By now, while the Army itself was experiencing the Duke of York's reforms, young blood was beginning to flow through the veins of the Corps. Names later to become famous, such as Fletcher and Burgoyne with Charles Pasley and John T. Jones, now begin to appear. Commissioned in the mid- to late-1790s, they and others who were not to survive the wars, communicated regularly with one another and grew into a pressure group that led to reforms taking effect towards the end of the Peninsular War.

Pasley and others were present at the battle of Maida (1806) where British infantry outfought Napoleon's veterans. The remarkably successful siege of Scylla followed, in which the Royal Engineers led by John T. Jones and others displayed their potential. More experience was gained in the disastrous expedition to Buenos Aires (1807)

Left: **Captain George Landmann** was the son of the Professor of Artillery and Fortification at the Royal Military Academy. He took part in the Vimeiro campaign and later received a commission as a lieutenant colonel in the Spanish Engineers. He was made a colonel in the Spanish infantry in which capacity he served at the siege of Matagorda. He wrote copiously about his time in the Peninsula in his *Adventures and Recollections*. The red jacket depicted was generally adopted in 1812. The blue normally worn by officers of the Ordnance service from 1782 could too easily be confused with the French. The Royal Military Artificers remained in blue until their conversion to Royal Sappers and Miners in 1813. The original of this uniform is the oldest held by the Royal Engineers Museum.

and in the attack on Copenhagen in the same year, after which the focus began to fall on Spain and Portugal when the French made their fateful attempt to absorb the Peninsula into their empire.

THE PENINSULAR WAR (1808–1814)

Although Engineers accompanied both Wellesley and Moore's campaigns in Portugal and Spain, in 1808 and 1809, and the Earl of Chatham's disastrous 1809 Walcheren expedition (see page 42), the engineering story really starts with the great plan for the defence of Portugal against the 1810 French invasion.

In October 1809, Lieutenant Colonel Richard Fletcher received his orders from Wellington for the great defence system north of Lisbon: the Lines

Left: The Royal Military Canal was the idea of Lieutenant Colonel Brown of the Royal Staff Corps, a former Royal Engineer, whose men provided much of the labour for the construction. This started in 1804 with the threat of invasion from across the Channel. On the one hand the canal would provide a quick means of moving troops between Shorncliffe and Rye; on the other it would become an obstacle to an invading force attempting to land on that vulnerable section of coast.

of Torres Vedras, and other measures designed to frustrate the French invasion. These succeeded. The enemy was halted at the end of 1810 after the successful defensive battle of Bussaco and four harrowing months for the Portuguese people, and withdrew back to Spain. Wellington was now able to tackle the strongly fortified 'gateways' of Ciudad Rodrigo and Badajoz. At Ciudad Rodrigo, after only eleven days of intensive work, conducted in the freezing January of 1812, the assault took place, led in by the engineers. In this short time some two miles of trench had been dug and five batteries constructed under very accurate fire from the garrison. It had cost two officers killed and five wounded from the Royal Engineer force of nineteen.

Badajoz, with its more determined and enterprising garrison, was a very different proposition and had already resisted two earlier attempts. Three weeks of unremitting effort in ghastly weather were needed to achieve the necessary breaches. The assault was an epic of valiant resolve in the face of almost unimaginable horrors. The five separate columns were each led in by two engineers; three were killed and

three wounded. Wellington's despatch at last triggered the necessary action to provide for a properly trained force of sappers:

> The capture of Badajos affords as strong an instance of the gallantry of our troops as has ever been displayed. But I greatly hope that I shall never again be the instrument of putting them to such a test as they were put to last night. I assure your lordship that it is quite impossible to carry fortified places by *vive force* without incurring grave loss and being exposed to the chance of failure, unless the army should be provided with a sufficient trained corps of sappers and miners …

Lieutenant Colonel John Fox Burgoyne accompanied Wellington as commanding Royal Engineer on his 1812 campaign after the triumph of Salamanca and occupation of Madrid. (Fletcher was recovering from a wound.) Burgoyne conducted two minor sieges but the inadequate resources of both artillery and engineers prevented the capture of Burgos after a frustrating siege and by November the army was back in Portugal to prepare for the sensational

Above: Lieutenant Colonel Sir Richard Fletcher, Bart, KCH* was Chief Engineer in the Peninsula from 1808 to August 1813, when he was killed at the siege of San Sebastián. Commissioned in 1788, he had served in the West Indies and the Mediterranean theatre before going to Portugal. Fletcher created the Lines of Torres Vedras. He was responsible for all the major sieges of the war except Burgos, having been left behind at Badajoz to supervise its reconstruction while recovering from a serious wound. Some felt that his amiable personality was insufficiently forceful to gain the staff support that the Engineers deserved, but his dedication and personal commitment was much lamented after his death.

*Knight Commander of the Royal Guelphic Order of Hanover. (See also page 248)

Above: The Lines of Torres Vedras, the bedrock of Wellington's plan for the defence of Portugal in 1810. Over a period of ten months 182 redoubts were constructed mounting more than 600 guns and with a manning capacity of some 40,000. Twenty Royal Engineers had been engaged in the work, although no more than eleven at any one time. They were joined by two engineers from the King's German Legion and four from the Portuguese Engineers. Eighteen Royal Military Artificers were allocated along with 150 soldiers from line regiments. Some 5,000–7,000 Portuguese labourers could be found working on the Lines at any one time. (Landmann's *Adventures and Recollections* RE Library)

BARROSSA (FEBRUARY 1811)

Sir Thomas Graham rode up to Birch and Nicholas after the action and took one in each hand, thanked them in the highest terms, and said, 'You are fine fellows at work as Engineers, you are fine fellows in the field, and I am more in your debt than anybody's.' This was reported in a letter from Captain Pitts to his father after the battle of Barrosa in 1811 in which several Royal Engineers and Royal Military Artificers fought as infantry. Graham had broken out from the siege of Cadiz in the hope of outflanking the besieging French army in a joint operation with a Spanish force. The latter failed to comply with his expectations when their route was blocked and the British contingent won an astonishing victory against the odds. (*Corps History* vol. I, quoting Pitts's private letters to his father)

TARIFA (JANUARY 1812)

(*Where the commanding Royal Engineer, Captain Charles Smith persuaded his reluctant brigade commander to stand firm*) To the British Engineer … belongs the praise of this splendid action. He perceived all the resources of the place, and with equal firmness and talent developed them, notwithstanding the opposition of his superiors; he induced the enemy … to open his trenches on the east, where, under the appearance of weakness, was concentrated all strength; finally, he repressed despondency where he failed to infuse confidence. (Napier's *History of the Peninsular War*, vol. 4)

1813 campaign that was to sweep the Allied armies to victory at Vitoria. Mobility was now the main consideration for the engineers deployed with their divisions, and a pontoon train was formed as a back-up. This soon proved its worth as the bulk of the Allied army debouched from Portugal through the rugged northern terrain south of Braganza and across the flood-swollen River Esla. The French armies were thus outflanked all the way to their fate at Vitoria.

A further year of fighting lay ahead. Wellington's Pyrenees campaign that ended in victory at the battle of Toulouse contained two great engineering feats. San Sebastián, the last serious resistance in northern Spain, was taken after a tough siege. At last there was sufficient artillery but the situation of the fortress and the determination of the garrison made it a bloody affair. However, it heralded a new age in the progress of the Corps with the appearance of the Royal Sappers and Miners, formed from the Royal Military Artificers under a Royal Warrant of August 1812 including the newly trained company of 'Pasley's Cadets' from Chatham.

BADAJOZ

Badajoz fell to Wellington's army on 6 April (Easter Day) 1812 at the third attempt. The five separate columns were each led in by two engineers. In all the attempts on Badajoz one-half of the Royal Engineer strength employed became casualties, half of these being killed. Twenty-four Royal Engineers and 115 Royal Military Artificers had been employed on this siege.

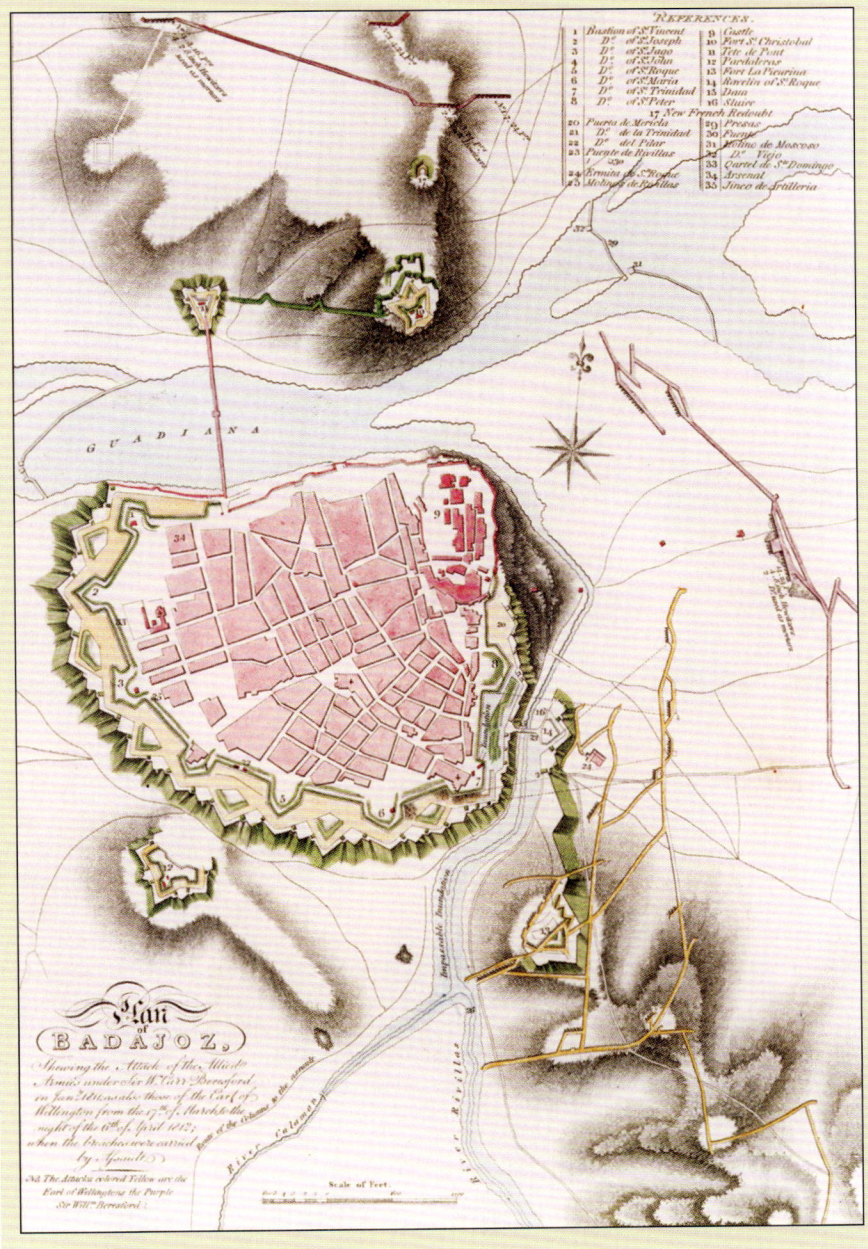

Above: The bridge at Cabezon being prepared for demolition during Wellington's retreat from Burgos in October 1812. This watercolour was painted by Captain (later General Sir) Charles Ellicombe and is one of several fine examples of his skill as an artist held in the RE Museum collection.

UNIFORM

Until military rank was granted in 1757, the engineers are assumed to have worn the same blue jacket as the artillery. They then appear to have adopted a distinctive red jacket until 1782, when the earliest surviving dress regulations for the Corps were published. The three portraits above illustrate the variations. Captain John Romer (left) is shown in 1710 in the long blue frock coat typical of the period. Romer also held a commission in 4th Foot with whom he served in the battles of Falkirk and Culloden, where he was badly wounded. Ensign Gother Mann (right) appears in 1763 in the red jacket with blue lapels and facings, the reverse of the artillery jacket. Gother Mann served in the West Indies in the American War and in Holland in the early stages of the Revolutionary Wars (1793), eventually becoming Inspector General of Fortifications, which post he held in 1815. Captain William Fyers (centre), pictured in 1790, wears the blue jacket that was standard throughout the Peninsular War until 1812 when scarlet was authorised. Fyers's uniform is described as a blue coat faced with black velvet lined with white, with white waistcoat and breeches. Fyers took part in the New York campaign of the American War and was at Walcheren in 1809. In between he was stationed at Gibraltar where this portrait was probably made. (*Corps History*, vol. 1, 1889 edition. The Fyers portrait is based on a miniature held in the RE Museum)

Above: The Siege of San Sebastián was the last great siege of the Peninsular War. Three-quarters surrounded by sea and protected by massive walls, the fortifications were breached so as to allow the assaulting troops to approach at low tide. Far more artillery was available than in the earlier sieges but, even so, the first attempt on 25 July 1813 was beaten back by a determined garrison. The second attempt on 31 August seemed to be suffering the same fate when General Sir Thomas Graham, who was in command, ordered the artillery to fire directly over the heads of the assaulting troops. This early example of a creeping barrage turned the day. The picture represents the right-flanking second assault. A diversionary frontal attack was supported by a massive mine laid in a drain that had been discovered leading through the walls. That attack, by the Royal Scots, was led in by Lieutenant Harry Jones, brother of Lieutenant Colonel John T. Jones (the chronicler of the engineer work in the war). Although the outer fortifications were captured there remained the castle, which only finally surrendered on 8 September. Harry Jones had been badly wounded and taken prisoner with several other soldiers. They suffered terrible hardships, having to lie out in the open with no medical treatment and endure the effects of their own side's bombardment during this time. Harry Jones eventually commanded the engineer force in the Crimea (see page 46).

THE ADOUR AND ALCÁNTARA CROSSINGS

To guarantee access to the north bank of the Adour at Bayonne in February 1814, Wellington ordered the construction of a bridge of boats below the citadel. Twenty-six local fishing craft (*chasse-marées*: two-masted vessels up to 50 feet in length) were required to cover the 260-yard gap. They were brought in under Royal Navy arrangements with sappers on board each one and anchored in place. The roadbearers comprised five massive cables carrying decking of three-inch planks. The anchorages were made from cannon sunk into the ground and a windlass system on the home bank took care of the considerable tidal movement. Gunboats and booms upstream protected the bridge from

hostile action. The bridge remained in constant use until it was dismantled on 18 May 1814.

This vast undertaking, described by Napier as '… one of the prodigies of war …' involved the Royal Navy, the Royal Staff Corps, the Royal Engineers and Royal Sappers and Miners. The timing and coordination of all the arrangements was critical as was the seamanship and discipline of both the Navy and the Sappers. The weather was bad and several boats were lost and their crews drowned. Captain Penrose RN, in charge of the naval side of things, said in his despatch: '… that so many chassemarées ventured the experiment, I attribute to there having been one or more sappers placed in each of them and a captain and eight lieutenants of Engineers commanding them in divisions. The zeal and science of these officers triumphed over the difficulties of navigation and I trust that none of their valuable lives have fallen a sacrifice to their spirited exertions.' Two of the sappers were drowned.

The Adour bridge was designed by Major Tod of the Royal Staff Corps.

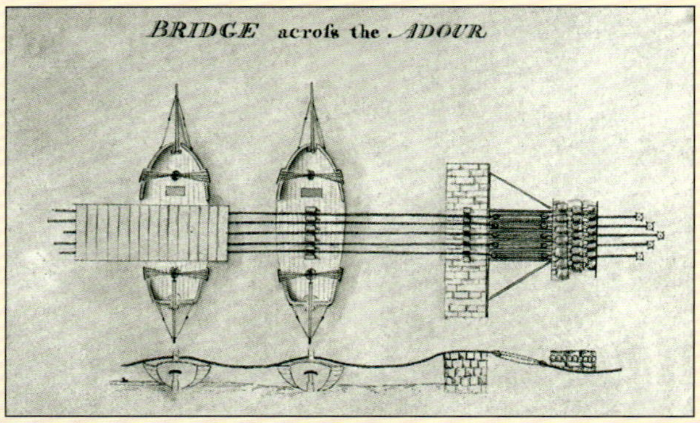

BRIDGE across the ADOUR

Similar principles, using tensioned cables, had been used earlier in the ingenious removable bridge to close the gap blown in the Roman bridge across the Tagus at Alcántara, thus allowing for reinforcement had this been necessary while at the same time giving Wellington a secure flank during his advance to Madrid in 1812.

Above: Major General Sir John T. Jones, Bart, KCB (1783–1843) was Fletcher's Adjutant during the construction of the Lines of Torres Vedras. He wrote the definitive works on the Lines and the sieges of the Peninsular War. He had already participated in many of the earlier campaigns of the Napoleonic Wars and took an active part in the subsequent operations until he was badly wounded at the siege of Burgos. He started writing his accounts of the sieges while recuperating. His criticisms of the Board of Ordnance so annoyed that body that they did their best to bring his career to an early conclusion. However, the Duke of Wellington, who held Jones in high regard, overruled the Board.

This change of name operated like magic. Everyone in an instant saw the propriety, nay, the absolute necessity, of the whole body being instructed in sapping and mining, and an institution was created by Lord Mulgrave for that purpose at Chatham.

In this siege Sir Richard Fletcher (he had been made a baronet for his earlier services) was killed. His replacement, Lieutenant Colonel Howard Elphinstone, now saw the Engineers through the last stages of the war. The pontoon train again performed sterling service in the advance across the Bidassoa into France and subsequent breakout in the battles of the Nivelle and Nive. But unquestionably the greatest engineering achievement of this period was the bridge of boats across the Adour downstream of Bayonne built to guarantee the blockade of that significant centre of resistance. Local fishing boats (*chasse-marées*) were hired on which to carry a roadway made from rope cables and timber planks. The bridge was designed by Colonel Sturgeon and Major Tod of the Royal Staff Corps and built by the Royal Engineers and Royal Sappers and Miners, whose officers and men had controlled the locally manned vessels into the river under Royal Navy command.

WATERLOO

Eleven Royal Engineer officers were present at the battle of Waterloo, acting largely in a staff capacity. The main interest of this campaign lies in the story of the Waterloo map (see box). Five companies of the Royal Sappers and Miners were

---

MEDALS FOR THE REVOLUTIONARY AND NAPOLEONIC WARS.

Four different awards were created for service in the Revolutionary and Napoleonic Wars (1793–1815).

The Army Gold Cross Described as a 'cross pattee in gold', it was awarded to general officers whose number of actions exceeded three. Actions are recorded on the arms of the cross and by gold clasps on the ribbon. No gold crosses were issued after the Peninsular War. Illustrated is the cross awarded to the future Field Marshal John Fox Burgoyne for actions at Badajoz, Salamanca, Vitoria, San Sebastián and the Nive.

The Army Gold Medal, set in a gold frame glazed either side, was awarded to field officers only. The obverse is a seated figure of Britannia. Actions are recorded on the reverse and by bars on the ribbon. The medal illustrated is also Burgoyne's, with clasps 'Salamanca' and 'S Sebastian'.

The Waterloo Medal was the first campaign medal to be issued by a British Government to all officers and men present at a battle (in this case, including the battles at Ligny and Quatre Bras). It depicts on the obverse the laureate head of the Prince Regent, and on the reverse, the seated, winged figure of Victory above the word 'Waterloo' and above Victory the word 'Wellington'. That illustrated is an example of one where the original iron suspension ring and clip have been replaced by a silver bar, common practice to overcome the problem of uniforms being stained by rust.

The Military General Service Medal was instituted only after the issue of the Waterloo Medal (see above) had caused an outcry from veterans of the campaigns since 1793 and their supporters. It took 34 years of argument before the government finally agreed to this silver medal depicting on the obverse, the diademed head of Queen Victoria and the date 1848; and on the reverse the Queen placing a laurel wreath on the head of the Duke of Wellington. Some 46 medals were awarded to officers of the Royal Engineers, 118 to the men of the Royal Sappers and Miners and 60 to members of the Royal Staff Corps. That illustrated has the clasps 'Nivelle', 'Ciudad Rodrigo' and 'Busaco'.

in theatre, of which two were with the pontoon train. None of the others was in a position to take part in the battle or the preparations for it.

Over the quarter century since the start of the wars, the Royal Engineers' strength had risen from 71 to 262. The Corps was effectively organised into 'battalions' of 48 officers each under a colonel commandant. The Corps was headed by the Inspector General of Fortifications.* The Royal Sappers and Miners' strength stood at 2,860. Inevitably peace brought reductions and

*The first incumbent was Lieutenant General Robert Morse, who took over on 1 May 1802 from General Sir William Green who had been fourteen years in the post then titled Chief Engineer.

by the 1820s the Royal Engineers were down to 193 with a half-pay list of officers yearning to be taken back on to the active list. The Royal Sappers and Miners amounted to 752 men, organised into twelve companies commanded by their own officers. By contrast, the Royal Staff Corps had risen to a peak strength in the 1820s of fifty-eight officers and 771 soldiers. They then declined rapidly and faded away in the 1830s.

However, after the defeat of Napoleon, the Royal Corps of Sappers and Miners, now properly organised and with proven talent, were to become fully stretched to meet the demands placed on them. They had come of age.

Above: Captain Marcus Waters in old age.

THE WATERLOO MAP

Wellington had visited the positions in front of Hal and Waterloo in 1814 and ordered his commanding Royal Engineer, Colonel Sir John Carmichael-Smyth, to prepare a map of the area. The southern portion of the map is dated around 12 June 1814 and the north-eastern 21 May 1815. Most of the north-western part was drawn by Major John Sperling RE 13–16 May 1815. When Wellington called for the map on 16 June 1815, a copy was not ready, one having already been supplied to the Prince of Orange. The assemblage of the original sketches had to be rushed from Brussels by Lieutenant Marcus Waters. Carrying the map fastened to his saddle, he was unhorsed at Quatre Bras, but recovered it when he found his horse eating vegetables in a garden. When Wellington returned to the Waterloo position on 17 June 1815, he again asked Carmichael-Smyth for the map. Wellington is said to have marked in pencil where he wished his army to be deployed and passed his map to his Quartermaster General, Sir William de Lancey, directing him to position the troops (marked with red arrow). De Lancey chose the crossroads near La Haye Sainte whereas Wellington preferred the Mont St-Jean position. Alternatively, Wellington may have marked the map when showing the Prussians where he wished them to join his army. De Lancey was wounded and carried off the battlefield. He was nursed by his wife Magdalene but died after nine days aged thirty-four. Major John Oldfield, Brigade Major RE at the battle, saved the map and wrote its history on it. It remained in the Carmichael-Smyth family until offered for sale in 1910 when it was purchased by Major William Harrison, Secretary of the Institution, to which he later presented it, and it is displayed in the Royal Engineers Museum.

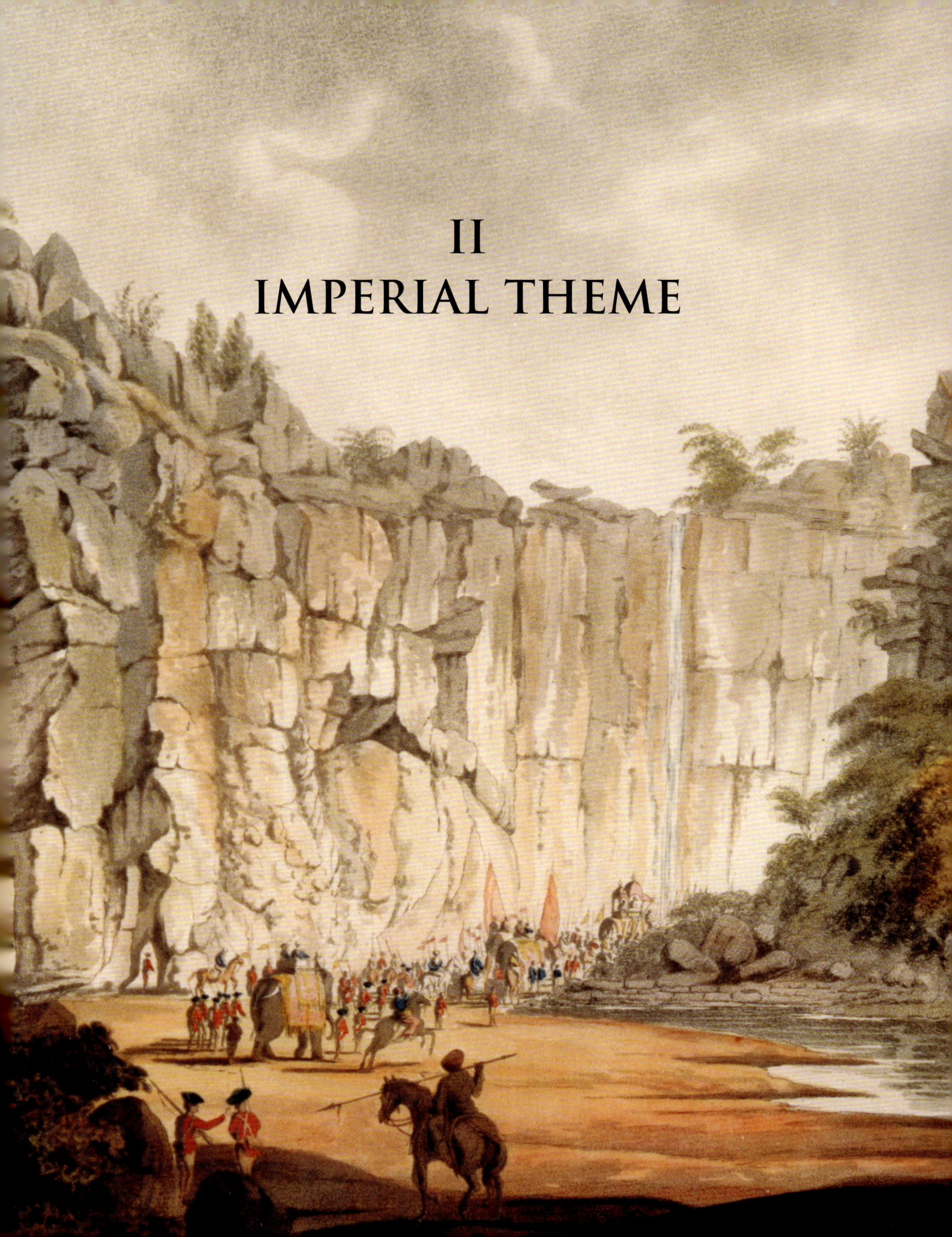

II
IMPERIAL THEME

6 | SCIENCE AND ZEAL

Above: Royal Engineer Officer 1812. This cut-out model shows the official dress when the jacket was changed from blue to scarlet. The aim had been to save confusion with the blue-jacketed enemy in the Peninsula.

As the men of the Plymouth Company of Royal Military Artificers wheeled into St Mary's Barracks at Chatham in May 1812, they must have been contemplating their future with mixed feelings. They were on their way to the war in Spain, from a comfortable billet as garrison troops, and were now due to be the first soldiers of their corps to receive special training in the arts of field engineering to prepare them for active soldiering. They had just experienced a year under the dynamic command of Captain Charles Pasley who had been sent to Plymouth to recover from severe wounds received in the siege of Flushing (1809–10). Pasley had already made plain his feelings on the 'inefficient and dangerous state of the [Engineer] Department' directly to Lord Mulgrave, the Master General of the Ordnance. At Plymouth he applied his energies and prodigious intellect to giving his soldiers grounding in their forthcoming tasks. He also studied new theories of learning and tested them out on his men.

The pressure to create a corps of sappers and miners came to a head with the training at Chatham, the creation of the Royal Corps of Sappers and Miners and the despatch of the first company in their new red jackets, dubbed 'Pasley's cadets', to join the siege of San Sebastián in August 1813. (See also pages 36–7.)

But the now Brevet Major Pasley's work had only just begun. The Royal Engineer Establishment soon formed in Brompton Barracks, was to develop under his influence into the heart of the Corps, the crucible in which the instrument was forged in terms both of organisation and training, that was to serve the army's needs for the whole of the nineteenth century and beyond. It was Pasley who recommended the concentration of training at Chatham as early as 1815 although that was not fully achieved until 1857 when the headquarters of the Corps moved down from Woolwich. By that time the Corps had taken over the whole of Brompton Barracks (1850). It was also Pasley

Above: General Sir Charles Pasley KCB (1780–1861) was the inspiration behind the early 19th-century reforms in the Corps, raising it from what he had described as 'damned and disgraced forever' in the eyes of the army to the position of envied repute it attained in his lifetime. He followed his career with a 'headlong ardour' from 1797 well beyond his retirement in 1846. His active service (Maida (southern Italy) 1806, Copenhagen 1807, Peninsula 1808–9, Walcheren 1809) ended at the siege of Flushing where he was badly wounded fighting like a 'desperate dog'. On recovery he was given the chance to prove his training theories at Chatham to prepare soldiers for the urgently needed role of fieldworks in the Peninsula. Thus were born the Royal Sappers and Miners and the Royal Engineer Establishment, now the Royal School of Military Engineering. In peace Pasley turned his energies into preparing the Corps for participation in the widest aspects of engineering, gaining a reputation for experimentation, innovation and achievement throughout the profession. He personally led these endeavours, taking a special interest in working with cements, and in diving and underwater demolitions. He was Director of the Royal Engineer Establishment from 1812 to 1846. He was elected a fellow of the Royal Society in 1816 and was Inspector of Railways from 1841 to 1846.

PONTOONING

Pasley designed a replacement for the 'tin pontoons' whose inadequacies had been exposed in the Peninsular War. His bipartite decked-in design could be carried on a two-wheeled cart, the two parts being joined at the stern on site to form a single pier or, with a second set, a bridge-supporting pier. Used in training until 1836, Pasley's pontoon was replaced by the Blanshard cylindrical design, a smaller variation of which went into service for infantry. Illustrated is one of the Blanshard models that were used for training purposes in the Model Room that was set up in the former chapel in North Block, Brompton Barracks after the Garrison Church was built.

Below: A cask raft rigged with a gyn for recovering submerged objects.

Below: Pontooning on the Medway in 1857; a water-colour by Second Corporal J. C. White.

Above: Simmons as a young man.

who had first advocated (in 1812) the integration of Royal Engineer officers with the ranks of the then Royal Military Artificers. Although this was not achieved formally until 1856, the officers and soldiers began to train together soon after Pasley arrived at Chatham.

The demand now was for instruction in the new technologies for civil and military applications at home and abroad. In 1825 the Master General of the Ordnance (the Duke of Wellington) had decided to do away with the Barrack Departments and to place the barracks of the whole United Kingdom under the Board of Ordnance. He added this responsibility to the duties of the Royal Engineers 'to the great advantage of the public service, and to a considerable saving' and decreed that 'practical architecture' should also be taught at Chatham. From this burgeoned the achievements covered in Section III of this book. Pasley produced a manual on the subject that remained a textbook for generations, being subsequently reprinted in 1862. His attitude, recorded in the preface, imbued all his teaching:

> The same kind of talents and knowledge that render an Engineer officer capable of forming the project of military works, and of arranging and directing great masses of workmen in a siege, or in fortifying the position of an army in time of war, by a very little more study will enable him to superintend the construction of public works of equal importance in time of peace ... By combining science and industry with the activity, zeal, and spirit of the military character, and by considering none of the multifarious duties that he may be required to perform as a drudgery, since they are all equally useful to his country, the young officer of Engineers has it in his power not only to establish a reputation for himself, but to contribute towards maintaining the fame of his Corps as one no less useful in peace than distinguished in war.

Left: Field Marshal Sir John Lintorn Simmons, GCB, GCMG (1821–1903) was gazetted field marshal in 1890. He was the third Director of the Royal Engineer Establishment, taking over from Sir Henry Harness in 1865. As a captain, he was appointed British Commissioner with the Turkish Army during the Crimean War, having placed himself at the disposal of the British Ambassador while on leave in the Near East when war broke out. The grasp of strategy and power of command that he displayed in this role gave his career a flying start and eventually led him to Lieutenant Governor of the Royal Military Academy, Inspector General of Fortifications and Governor of Malta.

SIEGE OPERATIONS included the detonation of a mine, during the 1837 siege operations conducted against the Chatham Lines. A series of these demonstrations, the last in 1877 until 1907 when interest in siegecraft was revived after the siege of Port Arthur in the Russo-Japanese war. (Watercolour by Captain C. E. Stanley)

Right: Siegecraft was an art more understood on the continent than in Britain. Engineers were sent to learn their skills abroad where expertise had been gained particularly in the long struggle between France and Spain for supremacy in the Spanish Netherlands. The science reached a peak in Louis XIV's France under his famous engineer, Sébastien de Vauban. The word 'sapper' and much more fortifications and siegecraft terminology derives from French. The basic principles were still being taught at 'The Shop' as illustrated in these 1815 drawings by Charles Beague.

A Survey School, established in 1833, took over that training from Ordnance Survey. When basic survey became a subject on the syllabus of the Royal Military Academy at Woolwich, an advanced course was introduced at Chatham. The Electrical School did not exist in its own right until after Pasley's time, but the Corps' introduction to the subject arose from his diving experiments (see page 112) which used 'galvanism' (electricity from a chemical cell) for blasting. The arrival of telegraphy in the Crimea added a dimension and for a time the 'Special Schools' at Chatham embraced these topics as well as submarine mining, chemistry, telegraphy, photography, photo-lithography and signalling. About the same time the Workshops were set up initially as support for the school and later for training tradesmen.

The Royal Engineer Establishment became the School of Military Engineering in 1869. This in no way diminished the tradition of creative experimentation that Charles Pasley had always advocated. His own work on hydraulic cement based on chalk and Medway clay made a substantial contribution to the development of artificial ('Portland') cement. Likewise his initiatives on diving broke new ground and laid the foundations of techniques for both military and civil applications. In both these fields Pasley made full use of, and gave full credit to, the men of the Royal Sappers and Miners whose part in the work went far beyond that of 'rude mechanicals'.

But the underlying purpose of the school remained its original one: training in fieldworks. The Fortification School developed from Pasley's 1812 fieldworks training and went on to embrace mining, bridging and demolitions. Trials in all aspects of this work were a major activity of the establishment. From 1830 onwards Pasley, who understood the value of good public relations, held an annual review to display the carrying-out of siege operations. According to Charles Dickens, in one of these, commanded by 'Colonel Bulder' (Pasley), Mr Pickwick was caught in an awkward predicament between the advancing lines of opposing troops. These displays continued into the 20th century.

Above and right: The Royal Engineer Institute, now the RSME Headquarters, was built in 1873 to accommodate the staff and work of what is now the Institution, including a museum and library. It also provided for teaching facilities, a lecture room, museum and some offices. It was designed by Lieutenant Ommanney, later Captain Sir Montagu Ommanney, GCMG, KCB, ISO. It exemplifies the Italianate style promoted by the Science and Art Department, based in South Kensington, in which Captain Francis Fowke, possibly the most distinguished of the Corps' architects, played an influential role. (See page 97.)

TRIALS were part of the business of the Royal Engineer Establishment, the results being immaculately recorded in the Records of Operations at the RE Establishment over the period 1822–66. These samples of the illustrations include a test that demonstrated the superiority of a simple bag of explosive over the more sophisticated manufactured petard for breaching the gate of a fortification.

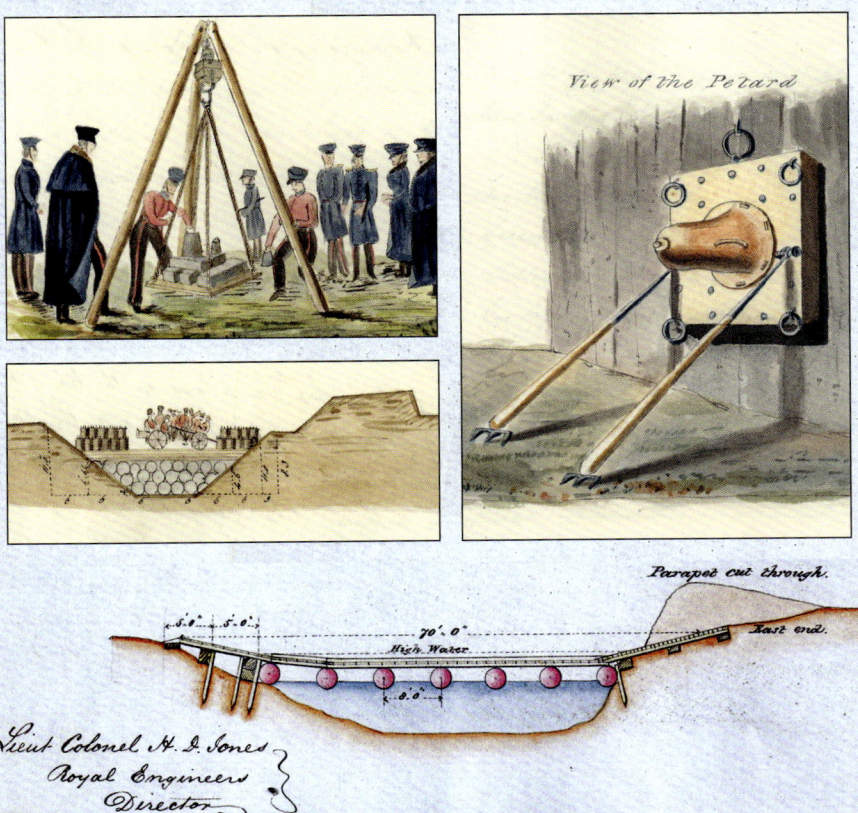

Above: The RE *Aide Memoire*, the essential vade-mecum for the aspiring sapper officer, was issued in two parts. Vol. I, 'formulae, tables, military statistics and memoranda for the field', contained handy advice on such matters as formulae for demolitions, laying out a camp and how to cure inflammation of the stomach and bowels in an elephant. Vol. II dealt with permanent engineering works and could allow the lonely Royal Engineer far from his base to build almost anything from a power station to a church incorporating Doric or Corinthian columns.

As a benign autumn gave way to bitter winter on the heights above Sevastopol in 1854, it became clear that the Allied armies were facing a protracted siege and a battle between opposing engineers. The first shots in this war of the alliance (Britain, France and Sardinia-Piedmont) against Russian expansionism had been fired by the Royal Navy in the Baltic. Later, a Russian offensive on the Danube had been turned back and there, in circumstances of utmost gallantry, Lieutenant James Burke, a Royal Engineer, became the first British soldier to be killed in action in the war. Sappers had been busy preparing field defences on the Gallipoli peninsula, to protect Constantinople from the perceived threat of a Russian attack, and converting the old Turkish barracks at Scutari to a hospital. In September the army had sailed to the Crimea from a cholera-ridden base at Varna in Bulgaria and gone into action for the first time since 1815. Three battles had been fought and won: the Alma en route to Sevastopol, Balaclava and Inkerman. The armies

Left: Lieutenant General Sir Harry Jones, GCB. The youngest brother of Major General Sir John T. Jones, author of the *Journal of the Sieges in Spain*. The two brothers served in the Peninsular War, Harry Jones being wounded and captured during the siege of San Sebastián, there having to endure the experience of being bombarded by his own side. His first command in the Crimean War was that of all the British Army troops that took part in the assault on Bomarsund in the Baltic campaign of 1854. He was the first Royal Engineer to be given a field command. In March 1855 he took over as Commanding Royal Engineer in the Crimea.

Below: The Attack on Bomarsund. Although known as the 'Crimean' War, the first action took place in the Baltic where Bomarsund in the Åland Islands was captured in 1854 after bombardment by a British fleet and an assault by an allied French and British force. In this first victory of the war, Brigadier General Harry Jones commanded the British contingent and Corporal Peter Leitch, later to win the VC in the Crimea, distinguished himself in the preparation of the batteries.

encamped on the heights and then only just survived the next assault, by a hurricane, whose ferocity caused far more casualties than had been so far sustained in battle. The losses at sea had dire consequences for the army. They included the steamship *Prince*, which carried the winter clothing and four sapper divers and their equipment. The *Rip Van Winkle* also went down with the two sapper photographers (Corporal John Pendered and Lance Corporal John Hammond) who, but for this disaster, would have been able to send back the first war photographs in history.

Above: Field Marshal Sir John Burgoyne, Bart, GCB. John Burgoyne was the son of 'Gentleman Johnny' Burgoyne of Saratoga notoriety. He was commissioned into the Corps in 1798 at the age of 16 and retired as a field marshal, the first sapper to reach that rank, in 1868. It was said that he had been under fire more than any other soldier in Europe. He survived unscathed the third and bloodiest assault on Badajoz, was wounded at San Sebastián, had his horse shot under him at Vitoria and took over temporarily as Chief Engineer when Sir Richard Fletcher was killed. After a spell in Ireland as Chairman of the Board of Public Works he became Inspector General of Fortifications and as such was a close adviser to Palmerston in the matter of the parlous state of the national defences, which was to bear fruit in the form of the 'Palmerston forts' many years later. He was over seventy years of age when he set out for the Crimea, as recorded in this chapter.

By April 1855 the sapper force had built up to some 40 Royal Engineer officers and four companies of Royal Sappers and Miners in an army of four weakened divisions. In March the Peninsula veteran, Major General Harry Jones, became Commanding Royal Engineer. His two predecessors had died, one of cholera and the other of sheer fatigue.

Above them had been the towering figure of General Sir John Burgoyne, now aged 72, special adviser to the commander, Lord Raglan. It was at Burgoyne's instigation that Raglan had concentrated on creating his firm base at Balaclava earlier in October. While this guaranteed the security of the British force it led to abandoning the prospect of a quick victory by *coup-de-main* assault on the city's defences, thus, in some opinions, condemning the army to the debilitating siege. Burgoyne had been the architect of the overall defensive plan and had undertaken difficult negotiations with the French allies on this matter. He returned to England in March to resume his job as Inspector General of Fortifications.

The maze of trenches grew as the months wore on. All depended on the sappers, for the construction of the batteries, the control of the infantry working parties and even just for guidance round the labyrinth of earthworks. The cry 'Follow the Sapper' took root here with particular poignancy when the infantry concentrated and moved forward to the assault. Bitter fights took place as either side made sorties into each other's works. In one of these, on 20 November 1854, the action took place that was to win the first Royal Engineer VC. Lieutenant Wilbraham Lennox, with a party of sappers, took part in a raid on some forward Russian rifle pits, converting them to home-side use. A similar affair on 9 April 1855, but on rather a larger scale, brought Colour Sergeant Henry McDonald his VC. He found himself in command of a large working party after the officers had become casualties, rallied the men as they were pulling out in the face of an enemy sortie, and led a charge that dispersed them.

Morale rose in the Allied camp with the spring of 1855. The bombardment took its toll of the enemy defences, only to be dashed by the

Above: The Baltic Medal (1854–5) (reverse), authorised for issue to the Royal Navy, was also awarded to those RE officers and men of No. 2 Company RS&M who were embarked to support the attacks on Sveaborg and Bomarsund.

Left: Balaclava, the base for all British operations, photographed by James Robertson.

Below: A general view of the Siege of Sevastopol drawn by Lieutenant (later Major General) John Cowell.

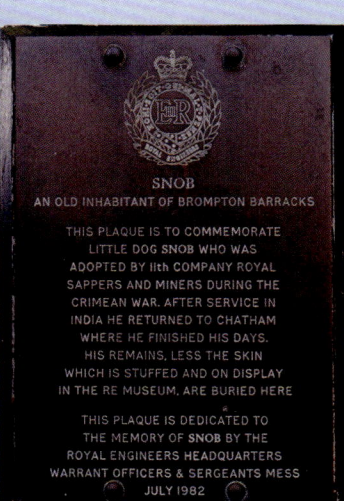

Snob, the dog that was befriended by soldiers, earning his name from his preference for the Officers' Mess. He was brought back to England, complete with campaign medal, but died as a result of an accident. His stuffed skin is on display in the RE Museum. His remains are buried in Brompton Barracks, close to the Crimea Arch.

SNOB
AN OLD INHABITANT OF BROMPTON BARRACKS

THIS PLAQUE IS TO COMMEMORATE
LITTLE DOG SNOB WHO WAS
ADOPTED BY 11th COMPANY ROYAL
SAPPERS AND MINERS DURING THE
CRIMEAN WAR. AFTER SERVICE IN
INDIA HE RETURNED TO CHATHAM
WHERE HE FINISHED HIS DAYS.
HIS REMAINS, LESS THE SKIN
WHICH IS STUFFED AND ON DISPLAY
IN THE RE MUSEUM, ARE BURIED HERE

THIS PLAQUE IS DEDICATED TO
THE MEMORY OF SNOB BY THE
ROYAL ENGINEERS HEADQUARTERS
WARRANT OFFICERS & SERGEANTS MESS
JULY 1982

Above: The Crimean War centrepiece. This beautiful piece is generally held to be the equal of any in the Mess collection. It was made by Stephen Smith, of Covent Garden, in 1867 and presented to the Mess by the Officers who served in the war against Russia 1854–6. The plinth carries silver plates with bas-reliefs of Crimean scenes: Fort Paul, Sevastopol and the interior of the Redan. The figures represent an RE officer, a private in the Royal Sappers and Miners in field service dress and a private in working kit.

ill-conceived assault on the Redan on 18 June the same year (the anniversary of Waterloo). The resultant carnage demonstrated the resilience of the enemy's defence, and the genius of their chief engineer Lieutenant Colonel Franz Todleben. The havoc wrought by the well-prepared Russians produced acts of extreme gallantry among the attackers and resulted in VCs for Lieutenants Gerald Graham and Howard Elphinstone, Colour Sergeant Peter Leitch and Sapper John Perie. The Allied bombardment continued, reducing the morale of the Russian defenders until on 8 September in a combined attack, the French won the Malakoff battery, the key point of the defence, but the British attack on the Redan had no more success than its predecessor in June. Events away from this immediate battlefield, however, had persuaded the Russians to abandon their resistance and withdraw from their battlements. Corporal Ross of the Royal Sappers and Miners made this discovery in the course of a one-man

Below: The Redan, photographed by James Robertson after the attack on 18 June 1855.

Above: View through an embrasure on the right of the 21-gun battery. This watercolour by Lieutenant E. R. James is reproduced from his journal, held in the RE Library. It was here that Corporals Lendrim and Wright climbed on to the roof, while under full enemy fire, to extinguish the burning sandbags, one of the actions that led to Lendrim's VC.

THE VICTORIA CROSS was instituted in 1856, inspired by the gallantry displayed in the Crimean War. It was to be awarded regardless of rank or race, for acts of valour performed in the face of the enemy and is now the highest award for gallantry. Since its inception, 1,358 awards have been made of which 55 have been awarded to men who have been members of the Corps during their military service. The special contribution of the officers of the Royal Engineers and the men of the Royal Sappers and Miners in the trenches before Sevastopol was recognised by eight VC awards. The group illustrated was awarded to Corporal John Ross for three separately cited incidents. The other medals are the Baltic Medal, the Crimea War Medal with Sevastopol clasp (with the unusual oak-leaf design), the French Louis Napoleon second issue *Médaille Militaire* and the Turkish Crimea Medal (French issue, the version intended for issue to the British having been lost at sea). The latter was instituted in 1852 and was awarded to a number of British servicemen for gallantry in the Crimean War.

Corporal John Ross had thirteen years of service behind him when he went to the Crimea. He was involved in many incidents in the hottest parts of the siege and had been an inspiration to his men working under fire as they pushed the trenches forward. He epitomises the achievements of the Corps in what had become a 'Sappers' War'. Ross continued his career after the war, eventually taking his discharge as a sergeant in 1867. **Above left:** the Ross VC (reverse).

Above: The Crimea Memorial Arch, recording the names of the 23 officers of the Royal Engineers, two Assistant Engineers and 161 men of the Royal Sappers and Miners who died in the war against Russia. The gates were cast from the bronze of captured Russian guns, the same material that is used for the Victoria Cross. The arch was designed by M. Digby Wyatt. The foundation stone was laid by the Commander-in-Chief, Field Marshal HRH The Duke of Cambridge, on 1 March 1860.

patrol he undertook with the object of rescuing one of his comrades after the failed assault (see illustration).

Across the Black Sea, the Allies were on the receiving end of the epic siege of Kars, the main Turkish outpost on their eastern frontier in present-day eastern Turkey. Lieutenant Colonel Henry Lake of the Madras Engineers was chief engineer in this heroic defence, which defeated repeated Russian attacks for four months.

At the end of the war, a further task awaited the sappers: the demolition of the principal military and naval installations of the base. By the time peace was signed in March 1856 the sappers had suffered 23 officers and 161 men killed. Never again would they go to war as two separate entities. The Royal Sappers and Miners, after 43 years of honourable service, became one with the Corps of Royal Engineers on 17 October 1856.

8 | SAPPERS IN THE RAJ

Below: The Fort at Vellore, where the sepoys of the garrison mutinied in July 1806, an event that presaged the Bengal mutinies fifty-one years later. One hundred and thirteen Europeans were massacred. Lieutenant Blakiston of the Madras Engineers took a prominent part in the suppression of the revolt. This watercolour is the work of General Sir James Caldwell GCB (1770–1863) who served in the Mysore wars during a distinguished career with the Madras Engineers, including the 1799 siege of Srirangapatnam. He was Senior Engineer, Mauritius in 1813 and Acting Chief Engineer, Madras in 1816. Caldwell was an accomplished artist and also designed St George's Cathedral and St Andrew's Church in Madras.

British India* owed much to military engineering. The East India Company could not prosper without their armies based on the settlements (eventually 'presidencies') of Madras, Bengal and Bombay; their armies soon discovered that they could not move or fight without their engineers and sappers. The security of the early trading settlements came to depend on the fortifications that grew up around them. The Company found it necessary to provide experienced military engineers to oversee all these works, such as John Call (q.v.).

Defence could not long remain passive. More active measures were soon necessary against both external and internal enemies. The defeat of the French in the Seven Years War (see page 26) ended the serious external threat for a century. Robert Clive's victory at Plassey (1757) had established British dominance in Bengal. More directly, Eyre Coote's defeat of the French at Wandiwash and Pondicherry, both dependent on successful siege work, effectively ended French power in India. By the 1770s the main threat to British interests was internal. A series of wars followed, essentially struggles for power

*Throughout this book the whole of the sub-continent is referred to as 'India' for all events up to 1947.

between Indian ethnic groupings in the outcome of which the Company had a strong commercial stake. These continued into the middle of the next century and, in the process, the boundaries of British India were set and the Company's army grew to meet the need.

The Mysore Wars (1780–99) against Haidar Ali and his son Tipu brought the Madras pioneers into being. They were recruited from line regiments and commanded by officers of the line who answered to the commander-in-chief but came under the Governor for tasking. This split command bedevilled the early days of military engineer organisation much as in the Queen's army (see page 51). A feature of these wars was the need to battle through mountainous jungle country to reach such objectives as the massive and remote fortresses of Nandidrug (formerly Nandidroog) and Savandrug (formerly Savandroog) to the north of Bangalore – quintessentially work for sappers. The 1799 siege and capture of Srirangapatnam (formerly Seringapatam) ended these wars.

The four Maratha Wars (between 1774 and 1819) brought the Bombay and Bengal Engineers into the action. They had formed in much the same manner and at much the same time as

those in Madras, a corps of engineer officers with a work-force of infantry-officered pioneers. In the early days the main campaigns were in southern India and therefore the Bengal Engineers were the least involved. These wars took place over vast tracts of difficult country. In general the aims were achieved and some brilliant work done. The Second Maratha War is notable for the parallel operations of Lieutenant General Gerard Lake's army operating from Cawnpore and Colonel Arthur Wellesley's from Mysore, the latter including the decisive victory at Assaye. Lake's equally successful campaign was spoiled by his attempted quick grab at the strongly held and extensive fortress of Bhurtpore (now Bharatpur) in 1805. Inadequate resources and bad appreciation led to disastrous defeat. Twenty-one years later Lord Combermere exacted retribution with an overwhelming all-Bengal army of 30,000 men with 112 siege guns. In this, the impeccable engineer plan depended much on mining. Lieutenant Colonel Thomas Anbury commanded the engineers. The final mine to be fired before the assault was charged with 10,000 pounds of gunpowder. Bhurtpore became the first battle honour of the Bengal Engineers.

The Nepal War (1814–15) was another 'internal' war following the pattern: armed raids from a neighbouring state – failed attempts at negotiation – retribution and annexation. It was remarkable for the determined resistance of the smaller enemy army and the new experience of

mountain warfare, engaging much effort from the Bengal Engineers and Pioneers.

By this time the Company's armies had already become embroiled in several expeditions overseas. All three engineer corps were represented in the first of these, to Egypt in support of Sir Ralph Abercromby's expedition in 1801. They were unable to arrive in time to participate but the seed was sown for a long and distinguished series of interventions by Indian troops in support of British interests, not least to the same country and Abyssinia in 1941. (See pages 70 and 82.)

About this time reorganisation in all three corps brought engineers and 'sappers and miners' under one overall command. Gradually the pioneers were absorbed into the sappers and miners resulting in a proficient overall body on which many demands fell as the influence of British India expanded. All three corps were involved in campaigns in Sind and Gwalior. In

General Sir John Call, who went to India in 1749 and was commissioned as a military engineer in the Madras Engineers in 1751. He was granted the rank of captain in 1759, just two years after the British Army engineers received the same privilege. Call raised temporary pioneer companies and commanded them both in the campaigns of the Seven Years War that ended the French presence in India and later against Haidar Ali. On his return to England in 1770, he was succeeded by Patrick Ross, one of only five officers of the British Corps of Engineers who served in India. This practice ceased in 1771.

Above: The Army of India Medal was similar to the Military General Service Medal (see page 39) in being issued retrospectively for earlier service. It was authorised in 1851 to those who had served in Indian campaigns between 1799 and 1826 and issued by the East India Company. The obverse carries the 'young head' effigy of Queen Victoria and the reverse the winged seated figure of Victory with palm tree and a trophy of arms. Only twenty-three of this very rare medal are known to have been issued to engineers. That illustrated, bearing the clasp 'Bhurtpoor', was awarded to Lieutenant (later Major General) William Forbes. He had to leave his work as architect of the Calcutta Silver Mint to take part in the siege of that city in 1826.

Right: General Sir Bindon Blood GCB, GCVO (1842–1940).
The first Chief Royal Engineer, he was appointed in 1936 at the age of ninety-three and continued until a month before his death at ninety-seven. Much of his early service was spent in India with the Bengal Sappers and Miners. He took part in the Zulu War with 30 Field Company (see page 78). Back in India he was Chief of Staff of the Chitral Relief Force and two years later in 1895 commanded the Malakand Field Force. In the South African War he commanded the Eastern Transvaal, as a lieutenant general under Kitchener, and returned to India for his last appointment as GOC Punjab Command.

the Sikh Wars, which finally resulted in the annexation of the Punjab in 1849, the engineer contribution came largely from the Bengal army. The climax of the first was the battle of Sobraon (1846) in which twenty-five Bengal Engineers took part along with eight companies of sappers and miners. The Second Sikh War was precipitated by a revolt in Multan, in the subsequent siege of which Major Robert Napier distinguished himself, as he did in the later actions including the decisive victory at Gujerat. His record in these wars and the Uprising was the springboard for the remarkable career that led to the rank of field marshal.

Then arrived the 'Devil's Wind', that watershed of all affairs in British India: the mutiny in the Bengal army and more general uprising of 1857 (see page 59). After it was all over the Company gave way to the Crown, the European element of the army became Queen's men and the regiments of the Company armies formed the new Indian Army, based initially on the Madras and Bombay units, which had remained loyal. The three corps of Engineers and the Sappers and Miners retained their former presidency titles. The officers remained on their separate lists for promotion but received exactly the same training as the Royal Engineers, through the Royal Military Academy, Woolwich and the School of Military engineering at Chatham. The former Indian Army cadet college at Addiscombe closed down.

After the Uprising, the Indian Army was again taken outside the boundaries of British India in the Second Afghan War and expeditions to China, Abyssinia and Burma (see pages 70 and 77). The North-West Frontier then became the principal concern. There a continuous sixty-year struggle for control of the border areas took

Left: The East India Company Memorial commemorates the absorption into the Corps of the engineer officers of the Indian service which, until 1862, meant permanent service in the army of the East India Company. This 'tazza' – rose-water container – was presented to the Corps by their brother officers, '… late of the Honourable East India Company's Engineers' in 1862. Medallion portraits of distinguished sapper officers appear on the main body while the plinth carries representations of the storming of the Kashmir Gate and the battle of Srirangapatnam.

Left: View within the Southern Entrance of Gundecotta Pass, an aquatint by Lieutenant (later Major General) Thomas Anbury of the Bengal Engineers.

Below: Indian Sappers and Miners. This painting by Johnny Jonas depicts an imagined scene in north-west India, with the soldiers of each corps shown in their distinctive dress. The Madras, Bengal and Bombay Sappers and Miners are respectively working on road construction (left foreground), a suspension bridge and a water supply point (right foreground).

THE THREE CORPS

In 1862 the Bengal, Bombay and Madras Engineers were amalgamated with the Royal Engineers, as part of the post-Mutiny reorganisation of the former Company's armies. The officers of the three corps in India remained on their separate lists for promotion. Addiscombe, the Indian Army's military academy, was closed. Officers were trained at Woolwich and Chatham as were all Royal Engineers. As the former 'Company's Engineers' retired or died, their places were taken by officers of the 'imperial' establishment. Such were the demands of the country that overall the engineer officer strength in India increased rapidly from 236 to 367 in the eleven years from 1854 to 1865. Many were employed on civil works and administration.

The Corps of Sappers and Miners were not greatly affected by the reorganisation. However, the supply of British non-commissioned officers was regularised by the formation of three Royal Engineer 'skeleton' holding companies in India, one for each presidency.

Top: Queen Victoria's Own Madras Sappers and Miners, 1896. l to r : Native officer and sapper in Review Order and sapper and colour-havildar in drill and marching order respectively. The Madras Corps originated in two companies of Pioneers, raised in 1780. They expanded to two battalions of which the 1st became the Corps of Madras Sappers and Miners in 1831, 'Queen's Own' in 1876 and 'Queen Victoria's Own' in 1923. Unlike the rest of the Indian Army the Madras Corps never recognized caste differences in Service matters. Their Depot was at Bangalore.

Centre: Officers of 3rd Field Company King George's Own Bengal Sappers and Miners, 1925. The commanding officer (seated third right) was Brevet Lieutenant Colonel Philip Neame VC DSO. The Bengal Corps was raised in the form of two pioneer companies in 1803 at Cawnpore. These produced the basis of the Corps of Bengal Sappers and Miners raised in 1819 at Allahabad by Major Thomas Anbury (see also page 55). They went through numerous changes of title until settling on the one above in 1923. The class composition laid down in 1907 was two companies of Hindustanis, Three companies divided equally between Sikhs and Punjabi-Mussalmans and one between Sikhs and Pathans. The Corps was based in Rurki (formerly Roorkee) from 1853 onwards.

Bottom: Some officers of No. 2 Company Royal Bombay Sappers and Miners, 1896. The Bombay Corps originated with a company of Pioneer Lascars in 1777. The pioneers had increased to four companies by the time of the Third Maratha War (1817). The Sappers and Miners formed in 1820 and became 'Royal' in 1903. In 1910 their composition was laid down as one-half Mussalmans (Punjabis), one-quarter Sikhs and one-quarter Marathas. Their depot was at Kirkee.

place, the direct consequence of the annexation of the Punjab. A complete reorganisation of the whole engineer command in 1885 led to greater appreciation of the part that sappers could play in support of these operations. In hundreds of expeditions, minor and major, all three corps were involved but inevitably the main responsibility fell to the Bengal sappers. Road and bridge building, field defences, demolitions (typically of village towers) became the stock-in-trade of all units.

Although the Indian Army was thus experienced in war by the beginning of the twentieth century, nothing could have adequately prepared them for the harsh conditions they were to encounter on the Western Front. Nevertheless, the three divisions sent to France in August 1914 effectively held the line at the critical time (see page 121). Outside Europe the Indian Army bore the brunt of the fighting in Mesopotamia (1915–18), and much of it in Palestine and East Africa as well.

Between the wars, affairs on the frontier were again the main concern and included the Third Afghan War (1919–20) (see page 65). Measures to pacify the frontier tribes are exemplified by the Waziristan Circular Road (see page 143).

The Indian Army again held the line for Britain at the start of the Second World War (see page 153), this time particularly in East Africa where the first of the three sapper VCs of the war was earned by Second Lieutenant Premindra Singh Bhagat of the Bombay Sappers and Miners. Thereafter, they contributed substantially to the Allied effort in the Western Desert and Italy and were the backbone of the Fourteenth Army in Burma.

The army that was handed over to India and Pakistan after the traumatic events of 1947 had to be ruthlessly dismembered into appropriate

CHITRAL

In March 1895 two expeditions set out to rescue the small, besieged garrison of Chitral, in the Hindu Kush mountains, 130 miles north of the Khyber Pass. The local tribesmen had earlier treacherously turned on a reinforcing party of Bengal Sappers and Miners and had taken Lieutenant John Fowler and others into captivity. Three companies of Bengal Sappers and Miners participated in the 15,000-strong Chitral Relief Force in which the Chief of Staff was Brigadier General Bindon Blood, later to become the first Chief Royal Engineer. Major Fenton Aylmer, who had already won the VC on an earlier expedition, was one of the 41 RE officers. Roads were built and the river crossings included a 90-foot span suspension built by Aylmer from timber and telegraph wire. Several actions were fought on the way but the enemy had dispersed when the relief force reached Chitral. It had been forestalled by another party under Colonel Kelly with a scratch force based on a group of pioneers with whom he had been working at Gilgit, 220 rugged and snow-covered miles to the east. Kelly was joined by a party of Kashmiri sappers under Lieutenant Leslie

Oldham. They battled their way over the Shandur Pass (12,230 feet), carrying the mountain guns when the mules failed, and suffering frostbite and snow-blindness. They fought their way through a strong enemy position, some of them having first to rope down a 250-foot cliff under fire. Sir George Robertson, who had so valiantly held Chitral for six weeks, wrote of Oldham:

'It was a rare deed for one British officer and twelve Kashmiri Sappers to cross that awful place to meet above them unknown ground and an unknown enemy; but the man who did it (an unsung hero) is a member of that famous Corps which gave Gilgit an Aylmer to blow open Nilt Gate, and a Fowler to cover himself and his comrades with glory even amidst the slaughter at Reshun.' (Lieutenant Colonel E. W. C. Sandes, *The Military Engineer in India*, vol I. p. 444)

This group of medals, awarded to **Subedar Major (Honorary Captain) Bir Singh** of the Royal Bombay Sappers and Miners contains three awards reserved for Indian Army servicemen. Bir Singh served on the Western Front 1914–15 and in Mesopotamia where he was decorated for gallantry during the unsuccessful attempt to relieve Kut in 1916. The **Order of British India (1st Class)** was instituted in 1837 by the East India Company for long and faithful service by Indian officers. It is a gold star with a crown and a blue enamelled centre bearing a lion. The **Indian Order of Merit**, also instituted in 1837 by the East India Company, is the oldest gallantry award of the British Empire. It is an eight-pointed silver star with a circular centre containing crossed sabres. The award ended in 1947. The **Durand Medal** (shown separately on the right), named after Sir Henry (see page 64), was originated in 1876 to be awarded annually for good and efficient service by a Viceroy's Commissioned Officer, NCO or sapper of any of the three Indian corps of Sappers and Miners. Since 1970 it has been awarded every three years to the Queen's Gurkha Engineers for the same purpose.

Also in the group, as well as the First World War and 1935 Silver Jubilee medals, is the **Indian General Service Medal**, of which there were four issues: 1854–1895, 1895–1902 (here illustrated, clasps 'Tirah' 'Samana' and 'Punjab'), 1908–1935 (here illustrated, clasp 'Afghanistan North-West Frontier'), and 1936–1939 (not illustrated). The reverse on all issues depicts the fort at Jamrud, which commands the Khyber Pass.

ethnic packages. Nevertheless, the soul of what had been built up over the previous two centuries survived in the shape of professional pride and tradition born of the exigencies of war. But the Indian Army had been more than a mere fighting machine. The needs of the population for good government and domestic well-being had led to a tradition of employing officers in political, professional and administrative roles outside regimental life. The fruits of this policy are touched on elsewhere in this book, but, as a result, there can be few countries that enjoyed such a close understanding between the civil and military communities.

Right: Winged Victory. This much-admired centrepiece was bought in 1920 for the RE Mess at Rurki as a 1914–1918 War Memorial. A reproduction of a statue found at Pompeii, it was made by a Naples silversmith, originally for another purpose, but its quality was recognised by the officers of the Bengal Sappers and Miners who then arranged for the mounting to be completed by a London artist. The panels represent the campaigns in France, Mesopotamia, the Indian Frontier and Palestine.

Widespread mutiny in India first flared up at Meerut (50 miles northeast of Delhi) on Sunday 10 May 1857. Firm action managed to confine the flames of revolt to northern and central India and to the Bengal army. They did then spread to the civil population, creating a more general uprising in those areas. Fortunately, the Bengal Sappers remained loyal, apart from two companies who sided with the mutineers early on, possibly moved more by panic than treachery, and served with distinction in subsequent operations. The threat to British interests was so serious that strong reinforcements were despatched from the Queen's Army. Speed was achieved by the lucky chance that an expedition to China was already under way and much of it was diverted to this more pressing business. Thus for the first time in history Royal Engineer units served in India alongside their Company counterparts.*

DELHI

The mutineers' first impulse was to concentrate at Delhi. A motley selection of troops was sent to contain them with the eventual aim of reoccupying the city. But the disparity in forces was too great, the objective too strong, disease too rife and leadership too weak for the British to do little more than hang on for several weeks. The situation became critical as elsewhere the most terrifying events were being enacted against the British popula-

*Including 4, 11, 21 and 23 Field Companies.

tion. The future of British India looked bleak. Things picked up in early July when Lieutenant Colonel Richard Baird Smith arrived to take command of the engineers. Although at that time there were some 15,000 trained Indian soldiers in the city to only 5,500 British, some enterprising young sapper officers had already worked out a plan by which the place could be attacked; Baird Smith was able to pull this plan together and convince the field commander of its success.

The most celebrated engineer event of the siege was the storming of the Kashmir Gate (see page 60). But the success of this and the other

Above: The Rajah of Futtegurh, one of the leaders of the Uprising.

Above: Captain Edward Fraser of the Bengal Engineers was the first fatal engineer casualty of the Mutiny. He was killed by a shot in the back fired by an Afghan sapper on 16 May 1857 at Meerut. Fraser had been ordered there with six companies of the Bengal Sappers and Miners to help douse the flames of the Mutiny that had flared up a week earlier.

Right: Colonel Richard Baird Smith, CB, was Chief Engineer for the siege of Delhi, the most decisive event in the suppression of the 1857 mutiny of the Bengal Army. Baird-Smith, ably supported by an outstanding team of Bengal Engineers, concocted the bold plan and convinced a vacillating commander to pursue it. Earlier in his career he had seen action in the Sikh Wars but then spent much of his time on civil works. On the outbreak of the Mutiny he had put the Rurki Garrison into an effective state of defence before being sent to Delhi. He never fully recovered from his exertions at Delhi, where he was badly wounded in the ankle, and died in 1861 at the age of 43. (Portrait as a cadet, RE Officers Mess)

LIEUTENANT ARTHUR MOFFAT LANG kept a journal throughout the Uprising. A modest man, whose ambitions lay in the field of engineering more than fighting, he displayed such bravery in the many actions in which he was involved that he was recommended several times for the VC. Not the least of his exploits was a daylight reconnaissance of the ditch at Delhi on the day before the assault under the noses of the enemy. He never received any reward for his services until the fiftieth anniversary, when a case was made that resulted in a CB. Extracts from his journal show his developing emotions:

Excited anticipation

18 May 1857. Truly an exciting time, but I find it suits me beautifully ... in fact the state of things, I am very happy to find, braces me up instead of depressing me or unstringing my nerves. The result shows me that a campaign would suit me, a fact I was not sure of, for I am very excitable on many subjects and fighting is one.

23 June 1857. This is no time for men to be idling and shirking; help is greatly wanted in Delhi, and I shall be much more useful there than I am here in inglorious ease and security.

In action

12 September 1857. This is splendid, no nonsense about it. We are fighting close up now, hurrying on the most rapid of sieges, working recklessly under fire. Our big smashing guns roar out together in salvoes and crash into the crumbling walls. It does one good to hear and watch a salvo of 24-pounder guns pounding the walls, and making the 'deadly breach'.

Witness to the death of Duncan Home VC

2 October 1857. Stevenson and I rushed up on the ruins of the bastion and saw Home run laughing up to the mine: he put his hand out and to our horror instantaneously the mine sprung; down we rushed, put every man to work to scrape and dig ... about 20 yards off in the hollow of a well I recognised his body, all mangled and covered with dust: poor fellow, his legs were broken into pieces, his arms broken and one nearly torn off; his death must have been instantaneous.

Revenge

16 November 1857 ... such a sight of slaughter I never saw. In the open rooms right and left of the archway Pandies* were shot down and bayoneted in heaps, three or four deep ... it was a glorious sight to see the mass of bodies, dead and wounded, when we did get in. [Lucknow]

After the death of his friend

25 March 1858. The death of my dear friend Elliot [Brownlow] has of course cast a gloom over the campaign as far as it concerns me, and has rendered it distasteful to me, and spoiled all my pleasure in war and victory ... He was a friend such as I can never find again in this world, more than a brother to me.

Right: Lieutenant Arthur Moffat Lang

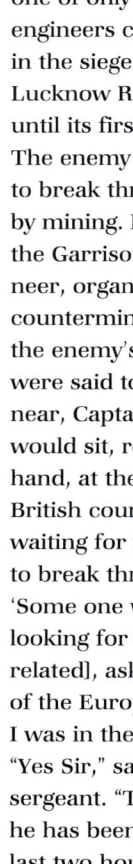

*Indian mutineers. The name caught on after the execution of the first, Sepoy Mangal Pandy.

Right: Captain George Fulton was one of only five engineers caught in the siege of the Lucknow Residency until its first relief. The enemy tried to break through by mining. Fulton, the Garrison Engineer, organised countermines. If the enemy's mines were said to be near, Captain Fulton would sit, revolver in hand, at the end of a British countermine waiting for them to break through. 'Some one who was looking for me [he related], asked one of the Europeans if I was in the mine. "Yes Sir," said the sergeant. "There he has been for the last two hours, like a terrier at a rat hole, and not likely to leave it all day."'

Left: The Kashmir Gate at Delhi was the objective of one of the four assaulting columns in the 1857 siege. It was breached by a party under Lieutenant Duncan Home of the Bengal Engineers, with Lieutenant Philip Salkeld, Sergeants Carmichael, Burgess and Smith, fourteen Bengal Sappers, ten Punjab Sappers* and Bugler Hawthorne of the 52nd Regiment. In the dawn light they covered the 200 yards to their objective under intense enemy fire, placed and fired their charges and signalled the assault party of the 52nd Regiment forward. The casualties included Carmichael, Smith and Sepoy Ram Heth (killed) and Salkeld and Jemadar Tillok Singh (mortally wounded). VCs were awarded to Home, Salkeld, Smith and Hawthorne. The Indian Order of Merit (1st Class) was awarded to Subedar Tula Ram. Two IOMs (2nd Class) and seven 3rd Class were also awarded. Duncan Home was accidentally killed only a few weeks later while blowing up an abandoned fort near Bolandshar, some 60 miles from Delhi (see Lang's diary above). This photograph was taken a few months after the battle. Other than the repaired bridge, the gate is much as it was at the time, with the right-hand portico filled in with brickwork.

*Five companies had been raised under this name by Duncan Home, from parties working on the Grand Trunk Road, before he himself went to Delhi from the Punjab.

Fulton was killed by a round shot but not before he earned the title 'Defender of Lucknow' for his work under ground and for the manner in which he had cheered the whole community by his optimism, perseverance, chivalry and daring.

three assault columns owed everything to the detailed knowledge of the ground acquired by meticulous but highly dangerous reconnaissance carried out by the junior engineers, such as Alexander Taylor and Arthur Moffat Lang. The climax of their work was the construction almost overnight of the batteries in no-man's-land that produced the breaches through which the columns attacked. It was a further seven days before the whole city was subdued. Three VCs were awarded to the engineers in the Kashmir gate party (q.v.). A fourth went to Lieutenant Edward Thackeray for his gallantry during the operations within the city.

LUCKNOW

On almost the same day as the fighting in Delhi subsided, 250 miles to the south-east the beleaguered garrison at Lucknow, who had held out in the Residency for four months against considerable odds, welcomed the first relief force under Generals Havelock and Outram on 25 September. Sir Henry Lawrence, under whose orders the preparations for the protection of the civilians had been made, had been killed by a roundshot, and among the other dead were two of the five engineers. They had achieved much by way of demolitions, preparing defences and countermining. Thus reinforced, the garrison

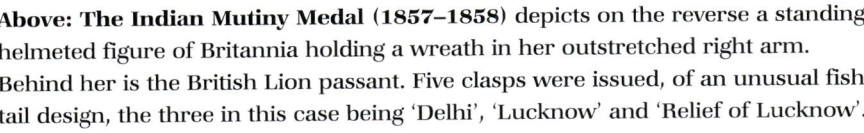

Above: The Indian Mutiny Medal (1857–1858) depicts on the reverse a standing helmeted figure of Britannia holding a wreath in her outstretched right arm. Behind her is the British Lion passant. Five clasps were issued, of an unusual fishtail design, the three in this case being 'Delhi', 'Lucknow' and 'Relief of Lucknow'.

was capable of a more aggressive defence but no breakout was possible for a further eight weeks until Sir Colin Campbell's army had fought its way through by 18 November. The Residency was then evacuated and Campbell's force escorted the women and children to safety, leaving a division behind under Outram to hold on outside the city which they were obliged to do for a further three months. Not until then had the situation elsewhere been brought under control and sufficient reinforcements built up from England.

These included two companies of Royal Engineers under Lieutenant Colonel Henry Harness, one of Madras Sappers and numerous pioneers. Of the engineers, many of those who had distinguished themselves at Delhi joined the force and all together, 1,800-strong, these formed an engineer brigade under Brigadier Robert Napier. He had been wounded in November but on recovery had joined Outram in February. The final battle for Lucknow began in the first week of March 1858. Much was demanded of the engineer force by way of reconnaissance under fire, battery construction, obstacle clearance and demolitions in the next three weeks until the rebel defence finally came to a bloody end.

CENTRAL INDIA

The epics underway in Delhi and Lucknow were matched in intensity by the more mobile operations farther south in Central India. First, any danger of insurrection in sympathy with the rebels in Oudh and around Delhi had to be nipped in the bud. Colonel Henry Durand of Ghazni fame (see pages 63 and 64), now in the political service, led these first moves with distinction with the help of a column from the Bombay army, including a sapper company. Both the Bombay and Madras armies had remained loyal and contributed greatly to these operations. Then Sir Hugh Rose arrived to take command of the Central India Field Force comprising only two brigades, including a company of the Madras Sappers, two from the Bombay army and, later, one of the Royal Engineers. In five months and in tremendous heat this force '... traversed 1,085 miles, crossed numerous large

rivers, took upwards of 150 pieces of artillery, one entrenched camp, two fortified cities and two fortresses all strongly defended, fought sixteen actions, captured twenty forts; and never sustained a check against the most warlike and determined enemy, led by the most capable commanders then to be found in any part of India.'

Brigadier Robert Napier joined the force in June as a brigade commander and shortly afterwards Sir Hugh Rose handed over full field command to him to conduct a final flourish of mobile operations into 1859 that proved his outstanding abilities in this role.

PEACE

Peace brought order, reorganisation (see page 54) and a change in relationships between Briton and Indian. Despite the sterling contribution of the Bombay and Madras Sappers and the Royal Engineers, the engineering had largely been a Bengal affair. More than half of the Bengal Engineers (64 officers) were engaged against the rebels, and of these more than half were killed or wounded. Eight VCs were awarded, four for Delhi, one for Lucknow, two for the Central India episode and one for subsequent operations.

Left: Jhansi was one of the chief points of resistance in the campaign in central India that was designed to distract rebel attention from the operations to relieve Lucknow while at the same time suppressing trouble in the central region. The Rani of Jhansi was a charismatic and effective military leader of her people who defended the city fiercely. The engineer force was reduced from eight officers to two during the siege, two being killed. Corporal Michael Sleavon of the 21st Field Company Royal Engineers was awarded the VC for his gallantry. The Rani escaped finally only to be killed in battle two months later. The map from which this drawing was taken was lithographed under the supervision of Major Howard Elphinstone VC, himself an accomplished artist, by the War Department Topographical Depot.

Left: Indian Mutiny Campaign silver. This piece commemorates the service of the officers of the Corps in India during the 1857 Uprising, the first time that Royal Engineers had fought in the sub-continent alongside their East India Company counterparts.

Above: The Afghanistan Medal 1878–1880 depicts on the reverse a martial scene of troops on the march with an elephant carrying a gun in the centre. Six clasps were authorised.

THE FIRST AND SECOND AFGHAN WARS
(1838–42 and 1878–80)

The two Afghan Wars share a reputation for military disaster born of political misjudgement. The Corps, however, emerged from both with its reputation enhanced.

Three episodes stand out from the First Afghan War in which military engineering turned crisis into accomplishment. But for the ingenuity and determination of the Bengal Engineers the main force, 'The Army of the Indus', would scarcely have started their long journey. A five-hundred-yard bridge over the Indus had to be conjured up from negligible resources available at the crossing site. Boats and timber had to be floated down 200 miles of river, rope had to be woven from grass, nails manufactured on site and

anchors devised from rock-filled wooden crates. Later, the progress of the whole force was in jeopardy before the fortress of Ghazni. A brilliant feat saved the day (see illustration). Finally, after a combination of political complacency, military ineptitude and Afghan treachery had led to the notorious retreat through the icy passes of the Hindu Kush and the death of 16,000 soldiers and their dependants, hopes for checking the tide of calamity lay in the defence of the fortified town of Jalalabad. There Captain George Broadfoot of the 34th Madras Infantry, who had earlier assumed the mantle of an engineer and raised six companies of 'Broadfoot's Sappers', had persuaded the commander of the garrison that a well-planned defence was the preferred option to retreat. His Herculean work confirmed his judgement. Jalalabad survived and became the base for the

Below: Ghazni, a strongly fortified city, obstructed the 1839 advance of the army escorting Shah Shuja, the new puppet ruler of Afghanistan whom the British wished to install on the throne in Kabul. A prolonged siege was impossible due to a dire shortage of supplies. A blockading force could not be spared or maintained, so by-passing was not an option nor was escalading the 70-foot walls. In a supreme act of heroism, the only gate was blown in. The party of one Bombay and two Bengal Engineers, three British NCOs and sixteen Bengal and six Bombay Sappers thus foreshadowed the Kashmir Gate affair at Delhi that was to win the Bengal Engineers three VCs. Lieutenant Henry Durand supervised the laying and firing of the charges. After a desperate fight at the gate, the city fell quickly to the infantry assault.

MAJOR GENERAL WILLIAM TREVOR, VC

Trevor was ten years old when his father, Captain R. S. Trevor of the 3rd Bengal Cavalry was murdered by the Afghans in the incident that precipitated the notorious 1842 retreat from Kabul. The surviving members of the family were taken into captivity for nine months. The new Afghan leader, Akbar Khan, used to amuse himself by setting up fights between young Trevor and the Afghan boys, the prize being a leg of mutton. William Trevor was commissioned into the Bengal Engineers in 1849. He saw action in the Second Burma War (1852–3) and the Uprising but was also engaged on survey and public works. He was awarded the VC for his part, with Lieutenant James Dundas, in assaulting a strongly defended blockhouse during the Bhutan War (1864–6).

Left: The Candahar, Ghuznee and Cabul Medal, awarded to the ten-year-old William Trevor for his experiences in the Second Afghan War.

Right: Major General Sir Henry Durand, KCSI, CB (1812–1871) first came to prominence as a lieutenant in the Bengal Engineers during the First Afghan War when he led the firing party at Ghazni. He was the offspring of a liaison, during the Peninsular War, between a British cavalry officer prisoner-of-war and a French lady. He spent nearly all his service in India, also seeing action in the Sikh Wars, but his remarkable political insight resulted in his holding many diplomatic positions, such as Commissioner in Tenasserim (1844) and Political Agent to Bhopal (1849) and Indore (1857). After the Uprising he was appointed Military Member to the Council of India. He then served for three years as Indian Foreign Secretary. A controversial character, somewhat overbearing in manner, who felt his services inadequately recognised, he was undoubtedly able and became something of an *éminence grise* to the Bengal Engineers. He died as the result of a fall from an elephant not long after taking up his final appointment as Lieutenant Governor of the Punjab. He is commemorated in the Institution's Durand Medal. (See page 58)

subsequent operations that somewhat rescued the pride of British arms.

By contrast, military operations in the Second Afghan War were well planned and executed, although arguably without sound political justification.

Although there was no engineering exploit so striking as the blowing in of the Kabul Gate at Ghazni in 1839, the engineers and sappers did excellent and useful work. They bridged the Kabul River many times, constructed dozens of fortified posts, built huts, laid telegraph lines and carried out the hundred-and-one jobs which help an army to advance and fight. Their surveyors covered huge tracts of unknown country, triangulating and mapping as they went and always liable to attack. And far behind the fighting men, the railway engineers toiled night and day to bring their lines through desert and gorge to the advanced bases.

The only real military disaster in the war was the total loss on 27 July 1880 at Maiwand of part of a brigade in a force despatched from Kandahar to meet an enemy threat from Herat. Lieutenant Thomas Rice Henn RE and a party of Bombay Sappers were among those killed. In a situation evoking Isandlwana the previous year (see page 80), Henn's name lives on for the gallant rearguard action he conducted before the entire force was wiped out. This led directly to General

Above: The Peiwar Kotal ridge was the scene of the first major battle on the central route of the invasion that launched the Second Afghan War (1878–80). Major General Frederick Roberts VC commanded this column, with Lieutenant Colonel Aeneas Perkins RE as Commanding Royal Engineer, which included only one small company of the Bengal Sappers. (From an album of watercolour sketches by Lieutenant Francis Longe in the RE Library)

Roberts's famous march from Kabul and the battle of Kandahar that brought the war effectively to an end. Some 167 officers and eighteen companies of sappers had taken part in the war. Five officers died and two VCs were awarded.

THE THIRD AFGHAN WAR (1919–20)

Before the Indian Army had recovered from the First World War, and against a background of civil disturbance in India, the Afghans decided to declare a 'jihad' and attack British territory in the area of the Khyber Pass. Fronts then developed from Chitral in the north to Baluchistan and even into eastern Persia. That in Waziristan grew into a full-scale uprising that was not really settled until 1925, although the war officially ended in the Khyber in May 1919 with a stalemate in Waziristan by the end of 1920 (see page 143). A consequence of this war was the construction of the Khyber Railway, a five-year military engineering project started in 1920 to provide reliable logistic support for future operations (see page 105).

Above: The Kabul to Kandahar Star 1880 commemorates General Roberts's famous march. The stars were made from guns captured from the army of Ayub Khan. The star illustrated was awarded to Colonel Aeneas Perkins, one of only twelve officers and two soldiers of the Corps to qualify.

Right: The Afghanistan Vase and Bowl commemorate the work of the Corps in the Second Afghan War (1878–80). The 24-inch-high vase and bowl are covered with 'beaten' decorations depicting Royal Engineers in action or on the line of march. This beautiful and unusual piece was made by a Peshawar jeweller named Ajhudia.

Below: The Sherpore Cantonment, outside the city boundary of Kabul, had been the home of the British occupation force, and its dependants during the First Afghan War (1839–42). Its lack of any defensive works contributed to the fateful decision to evacuate Kabul in January 1842. When the British returned in 1879, under the command of Major General Frederick Roberts VC, they found fortifications that had been developed by their old enemy, Dost Mohammed, and were able to improve them to guarantee the security of the army. They proved their worth against the mass attacks that were launched in December, the defeat of which was the essential first step to eventual victory. This photograph shows the Bengal Sappers and Miners guarding their sector.

Throughout the Empire, particularly in India, civil and military control could make little progress without maps. But the survey of India went far beyond this limited aim. The skills of the Survey Branch also contributed vastly to the science of geodesy and to topographical knowledge. All arms were represented in this work, not least William Lambton of the 33rd Infantry and George Everest of the Bengal Artillery. First in the field, however, was James Rennell, of the Bengal Engineers. Lambton launched the triangulation of the whole of India and from this Everest brought what then became the Great Trigonometrical Survey to its climax before handing over to his successor, the Bengal Engineer, Andrew Waugh. He joined the Survey in 1822, participated in Everest's measurement of the 'great arc of the meridian' from the southernmost point of the Indian peninsula to Dehra Dun in the north. During his time as Surveyor-General and

THE PERILS OF SURVEY
Captain James Rennell FRS, whose Bengal Atlas was published in 1777, reported thus after being attacked by a group of hostile fakirs:

One Stroke of the Sabre had cut my shoulder bone thro', and laid me open for nearly a foot down the back, wounding several ribs, besides a Stab in the same arm and a large cut in the hand, which has deprived me of the use of my forefinger … I must not forget to tell you that about a month ago a large Leopard jumped me, and I was fortunate enough to kill him by thrusting my Bayonet down his Throat. Five of my men were wounded by him, four of them very dangerously. You see I am a lucky Fellow at all times.

Superintendent of the Topographical Survey, he extended the triangulation to an area of some 316,000 square miles, equivalent to three times that of the British Isles.

The thirst for geographical information also played a crucial part in those frontier-probing

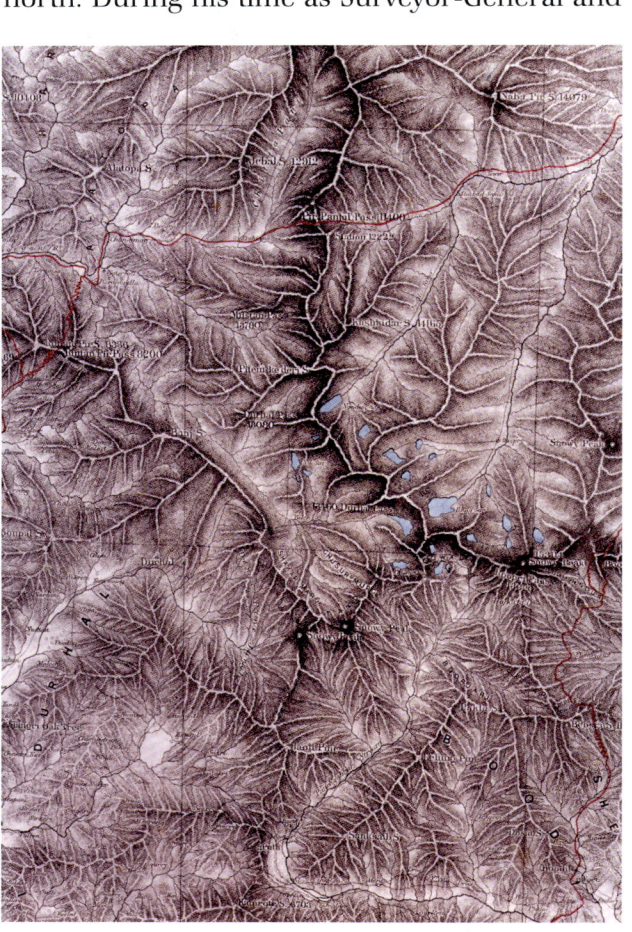

Left: Lieutenant Colonel T. G. Montgomerie FRS (1830–78), was a pioneer surveyor of the glacial regions, among his achievements being the discovery of the peak K2. He conducted the survey of Kashmir, a gruelling task that accounted for his early death. He is particularly noted for initiating the system of clandestine cross-border surveys by native explorers. His map of Kashmir (**far left**) provoked the following response from Lord Canning, the Viceroy: 'I never saw a more perfect or artistic-like production of its kind than this map … You have given me new proof of what I long discovered that there never was a more able, zealous, reliable body of English gentlemen brought together under any government than the Corps of Engineers of Bengal.' Montgomerie's name is commemorated in the Institution prize for contributions to Corps publications on professional subjects by junior officers and soldiers.

activities now known as the Great Game. The delicacy of international relations would not permit official survey parties to operate in most of the border regions such as in the Pamirs, in which the Russians had such a close interest, or in Nepal, let alone across the Himalayas in Tibet. However, knowledge of these areas was seen as essential to the security of India. The problem was overcome by Captain (later Lieutenant Colonel) Thomas Montgomerie who trained Indian specialists to enter these areas in disguise. There they would measure distance by carefully rehearsed pacing, and altitude by recording the boiling point of water. Bearings would be taken with a sextant that had to be concealed in the false bottom of a travelling chest. These incredibly brave men became known as the pundits. The best known was Nain Singh who was well rewarded for his efforts with the award of the CIE, a grant of land and money and the Royal Geographic Society's gold medal.

BOUNDARY COMMISSIONS

Later in the 19th century boundary surveys were in demand around India and farther afield. Frequently these were undertaken by joint

Two pictures from the collection of photographs taken on the 1858 North American Boundary Commission in the years 1860–1: (above) A camp on the 49th Parallel between Haig's Pond and Rock Creek and (below) Cutting and Stone Pyramid on the 49th Parallel at Kitheman looking west.

Right: The Waugh Epergne was presented to Major General Sir Andrew Scott Waugh KCB, FRS on his resignation of office and service in India in 1861. Waugh had taken over from Sir George Everest and was responsible for announcing the survey of the world's highest mountain and naming it after his predecessor, Sir George Everest.

international commissions to settle contested borders. The most distinguished practitioner in this matter was Thomas Holdich (later Colonel Sir Thomas, KCMG, KCIE, CB). With experience from the Abyssinian campaign and the Second Afghan War, he took part in the Russo-Afghan boundary commission (1884), undertook the demarcation of the India–Afghanistan boundary, the Durand Line (1894), the Pamir boundary commission (1895) and several others. After retirement he led the 1902 commission that settled the Chile–Argentine boundary dispute.

Africa, with the haphazard borders resulting from the 'scramble' was a rich field for such work. Corps History records twenty-two commissions on that continent covering some 15,000 miles between 1890 and 1911. The dangers included hostile action by local tribes. In the case of the 1907–9 commission on the borders of Nigeria and German Kamerun, the commission lost five killed and nineteen wounded in a two-day battle in which the British and German escorts fought side by side.

The most renowned commissions, however, were those that settled the North American boundary between Canada and the United States. The first, in 1843, took three sapper officers, two of whom were astronomers, and twenty soldiers of the Royal Sappers and Miners, to the New Brunswick area to settle points that had been outstanding from various treaties back to 1773 and to cut the boundary line. The difficulties of working in dense forest with no means of communication beyond birch-bark torches and horns can be imagined, but these were aggravated by having to work in temperatures down to minus thirty degrees at which point the chronometers frequently stopped and the thermometers failed to read. The second commission started in 1858 to deal with the section of the 49th Parallel from the west coast across the Rockies. Four sapper officers participated, including the British Commissioner, Colonel John Hawkins, together with fifty-six soldiers of the Corps. A twenty-foot-wide path was cut along the line of the parallel, marked with iron pillars at three-quarter-mile spacing. The work was completed in 1862 having carried on through two winters. It left a final gap across the prairies to join up with the early survey. This was accomplished by the 1872 commission working through unexplored territory, plagued by cold and storms in the winter and biting insects in the summer.

EXPLORATION

Exploration other than for survey was an activity indulged in by hundreds of sappers as the world opened up during the 19th century. Often these adventures were in pursuit of sport or some scientific interest such as archaeology, geology, botany or anthropology. Nearly always they also brought back valuable geographic information and many made outstanding contributions to science. Henry Morshead, who as a lieutenant had explored the Tsang-po with another officer on a 1,680-mile expedition in 1913, took part in the 1922 Everest attempt, reaching 25,000 feet before being defeated by frostbite and exhaustion. Sport, in the form of big-game shooting, took Lieutenant Harald Swayne to Somaliland in 1885. His knowledge of the country from this and future expeditions, recorded in his *Seventeen Trips through Somaliland*, was the only detailed record of the country at the time. It proved invaluable when trouble broke out in the late 1890s and several expeditions had to be mounted to restore law and order, the last, in 1903, against the 'Mad Mullah' almost on the scale of the Abyssinian campaign of 1867.

The survey of Palestine under the auspices of the Palestine Exploration Committee was very much more than a map-making operation. It started simply as a basis for providing a new water supply for Jerusalem. It developed into a detailed survey from Lebanon to Sinai in an attempt to identify the biblical sites and

Another rare award in the RE Museum collection is the **Zanzibar Order of the Brilliant Star**, awarded to Brigadier H. H. Austin, a surveyor and indefatigable explorer in many parts of Africa, particularly the Sudan.

COLONEL MARK BELL, VC, was the first recipient of the MacGregor Memorial Medal. At the time of the award he had already carried out a number of expeditions, bringing in valuable intelligence, to northern China (1882), south-western Persia (1884), Mesopotamia and Armenia (1885/6) and western China and Kashgaria (1887). The latter first brought the young Francis Younghusband, a junior member of Bell's staff, to public attention. Lord Curzon wrote of Bell: 'His extraordinary travels over almost the whole Asian continent ... entitle him to be considered the territorial Ulysses of this age.' Bell had won his VC in the 1873 Ashanti War, in a ferocious fight with the Ashanti rearguard, while leading the route-clearing party. (See page 78.)

The MacGregor Memorial Medal was founded in 1888 by the United Service Institution of India in memory of Major General Sir Charles MacGregor for 'the best military reconnaissance, exploratory journey, or survey in the remote parts of India or its neighbouring countries presently under the jurisdiction of India of the year'. MacGregor, a cavalryman, had himself undertaken numerous dangerous explorations around the borders of Afghanistan in the course of an active career, finishing as Quartermaster General in India during which appointment he concentrated particularly on bringing the Intelligence Department to a high pitch of perfection.

THE SURVEYS OF PALESTINE ran from 1865 to 1883. They followed a series of phases from the original survey of Jerusalem through a separate Sinai expedition under Charles Wilson to the main Western Palestine survey that ran from 1872 to 1878. Eastern Palestine was covered in 1881 and Kitchener himself returned in 1883 to clear up the last section of Sinai. The original Jerusalem survey had been financed by Miss (later Baroness) Burdett Coutts, the remainder by the Palestine Exploration Society whose aim was principally to identify sites mentioned in the Bible. The 1870 Sinai survey party is illustrated in two groups that respectfully separate officers from other ranks. Fortunately Colour Sergeant MacDonald, whose brilliant photographs demonstrate both artistic and technical skill of a high order, must have entrusted his precious camera to one of the officers for the second shot. The 1870 expedition comprised Captain Charles Wilson (leader, third from right with pipe), Captain Henry Palmer (who had worked in Canada under Moody (see page 102), Mr E. H. Palmer (Arabic scholar), Mr Wyatt (naturalist), the Rev N. W. Holland, Colour Sergeant James MacDonald, Corporals Brigley and Goodwin, Lance Corporal Malings from Ordnance Survey and guides Jemma, Hassan and Salem.

examine their related archaeology. The work lasted over the period 1865 to 1883. Several of the officers involved later rose to high rank, such as Charles Wilson* (see page 82) and Charles Warren (see page 88); and not least Herbert Kitchener who took part in 1874–7 and displayed courage and a cool head in the dangerous situations experienced by the survey parties. He also started to learn Arabic despite the arduousness of the work and the time he had to devote to the detail of the maps and reports.

*Charles Wilson had also been Secretary of the 1858 Boundary Commission in British Columbia.

Right: The Convent of St Katherine, Sinai, an example of Colour Sergeant MacDonald's photography.

12 | EMPIRE IN ACTION

While the army in India existed primarily to maintain and develop East India Company and British interests within India itself, it also became a base for overseas expeditions in the protection and promotion of Britain's wider concerns. Several small, often brutal, wars were fought to this end in which the Queen's Army also frequently provided an element.

THE BURMA WARS

Burma was a continuing thorn in the flesh. Border disagreements were the trouble particularly after the Burmese occupied the Arakan, whose native people sought succour from India. The First Burma War (1824–6) followed genuine efforts to achieve a peaceful solution. It was hoped that the war might be confined to the coastal areas because, as the Commander-in-Chief put it, if the army penetrated into the interior they would find 'nothing but jungle, pestilence and famine'. An 11,000-strong expedition set forth under Major General Sir Archibald Campbell with ten engineer officers and a battalion of the Madras Pioneers. Rangoon fell quickly but lack of transport and supplies prevented further advance. Only in the new year of 1825 was Campbell able to advance to threaten the Burmese king's heartland at Ava, near Mandalay, and then only with a small force supported by the Royal Navy on the Irrawaddy. Time and again it seemed that he would be forced to withdraw as the enemy either rallied to fight back or won time by bogus negotiations. Often the logistic problems seemed insuperable and men fell in large numbers from disease. Nor was there respite for the sappers in such terrain. In the end the superior discipline of the British forces won the day and a treaty was signed in February 1826.

Below right: General Sir Harry Prendergast, VC, GCB, was commander of the British forces in the Third Burma War (1885–6). His well-planned campaign led to the surrender of the Burmese army and the abdication of the king and queen in a matter of weeks, but four years of guerrilla war followed before Burma came fully under British control. Prendergast was a Madras engineer who had been awarded the VC for a very gallant rescue in the civil uprising associated with the Mutiny, after which he took a highly active part in the rest of the fighting. His frequent applications for further active service finally bore fruit in Sir Robert Napier's Abyssinian campaign in which he commanded three companies of Madras sappers and was present at the storming of Magdala. After Burma he served in the Indian political department as Resident at Baroda and Mysore.

Trophies of the Burma Wars. The bell was captured in Rangoon during the Second Burma War (1852–3) by the Bengal Engineers and erected on the Square at Brompton Barracks in c.1904. The cannon is one of two taken at Mandalay in the Third Burma War (1885–6) by the force under Sir Harry Prendergast. They were presented to the Corps by Lieutenant General George Sanford who had been the Commanding Royal Engineer in the force.

An uneasy peace lasted for a quarter of a century but gradually the Burmese made life increasingly difficult for the British merchants in the country. Pressures built up until a relatively minor incident sparked off the Second Burma War (1852–3). A thoroughly well-prepared and equipped expedition was mounted under Major General Godwin, a veteran of the first war, amounting eventually to two small 'divisions', one of which was commanded by the sapper, Brigadier General Sir John Cheape. Eighteen engineer officers and four companies of Madras Sappers and Miners made up a powerful sapper force. After numerous battles, mostly in the south of the country, in which engineering skills of fortification and siegework were fully exploited, the war ended suddenly with the deposition of the king. A treaty was signed with his successor under which the whole of the southern area of the country including Rangoon and the Delta were annexed under British rule. This gave the Burmese no outlet to the sea.

A new king, Thibaw, came to the throne in 1878, determined to end British sovereignty. Several serious incidents led to pressure from the British side for the annexation of the rest of the country, not least the public execution of eighty of his near relatives. Fears were reinforced by the growth of French influence with its threat to British trade interests. The pretext for the Third Burma

Above: The Indian General Service Medal with clasps for 'Burma 1887–89', 'Chin-Lushai 1889–90' and 'Burma 1889–92'.

War (1885–6) was soon enough established for a 9,000-strong expeditionary force to be launched up the Irrawaddy. Thirty-five sapper officers and six companies of Sappers and Miners took part. The force was commanded by Major General Sir Harry Prendergast, who had won the VC in the Mutiny. Within two weeks of setting out King Thibaw had surrendered and was on his way to exile. A difficult guerrilla war then started and continued for seven years before Burma was pacified under the British crown.

THE CHINA WARS

The bone of contention in China was trade rather than territory. The first of the China Wars, often called the Opium Wars, which ran from 1841 to 1843, was instigated after negotiated arrangements for trade in Canton had been interfered with. Joint operations with the navy, in which the engineer element was provided from the Madras army, took place up and down the coast of China. The capture of Canton by the army under Major General Sir Hugh Gough was the first prize. The Chinese were reluctant to negotiate, however, and the war was taken northwards with the eventual capture of Shanghai and Chinkiang, threatening Nanking. This resulted in the treaty of Nanking under which Hong Kong was ceded to Britain and trading rights established in certain ports.

Below: The Taku forts (see text) guarded the mouth of the Peiho River on the approach to Peking. Those on the northern bank were the objective of the attack on 21 August 1860, engineer reconnaissance having established that their capture would make untenable the southern forts, one of which is illustrated here. This remarkable photograph was taken by Lieutenant John Papillon and forms part of a collection that so impressed General Sir John Burgoyne that he passed it to the Prince Consort for his examination.

The non-compliance of the Chinese with these agreements brought on the second and third China Wars (1857–9 and 1860). A substantial force from the Queen's army despatched for the purpose had to be diverted to India to deal with the Mutiny, so it was not until mid-1858 that order could be restored in Canton, the centre of most of the Chinese aggression, and diplomatic efforts made to reach a permanent settlement, this in conjunction with the French,

MAJOR GENERAL CHARLES GORDON, CB, seen in this portrait by Val Prinsep in the regalia of a Chinese mandarin at the age of 34 while still a captain. He had come to public prominence as the enterprising commander of the irregular Chinese force known as 'The Ever Victorious Army' to which he had been transferred while in China in a more conventional role in the Third China War (1860). Returning to England in 1865 he came down to earth as CRE Gravesend, where apart from his professional responsibilities, he revelled in the opportunity to satisfy his strong Christian-based philanthropic convictions by setting up boys' clubs and helping the poor in other ways.

Thereafter Gordon's career again followed an unconventional pattern. In 1874 he began those links with the Sudan for which his name is mostly known; a three-year tour as Governor of the Equatorial Provinces (southern Sudan) trying to suppress the slave trade and in the process mapping and opening up vast new tracts of country, three more as Governor General (1877–8) and finally on his special mission to resolve the crisis that had arisen over the Mahdi's revolt. He was trapped in Khartoum and killed on 26 January 1885. (See page 83.)

Charles Gordon remains one of the most enigmatic figures of the 19th century. The inner strength that drove him to almost prodigious feats grew from his religious convictions. His scorn of death may have accounted for his fearlessness in battle. When it came, this and the selfless manner of his life inspired veneration in the widest public of his day.

Below: Tai Ping regalia brought back from China by Gordon after his sensational campaign in command of 'The Ever Victorious Army'. The Tai Pings were able to make this richly embroidered ceremonial dress, contrasting with the plainer imperial style (as in the jacket on the left), following their capture of the silk factories at Nanking. After their defeat the emperor ordered the destruction of all such symbols of office, so these are rare, possibly unique, survivals.

Bottom left: A letter from Gordon, found on the decapitated body of one of the Tai Ping leaders, trying to persuade them to surrender the city of Soochow peacefully in exchange for a guarantee of their safety and that of the citizens of the city. They eventually accepted the terms but the Chinese imperial leadership reneged on this promise. The principal leaders were executed and the city sacked, to Gordon's deep remorse and fury. He was only just persuaded not to resign from his command following this episode. The story is told in the book, also illustrated, E. A. Hake's *Events in the Chinese Rebellion.*

Above: The China Cup commemorates the 1857–60 war in China. The body of the cup is intricately carved with battle scenes in which English and Chinese troops are represented.

Above: The China War Medal 1856–60. The reverse shows a collection of war trophies with, in the exergue, an oval shield displaying the Royal Arms, all positioned beneath a palm tree. Above the words ARMIS EXPOSCERE PACEM, and in the exergue the word CHINA. Six clasps were authorised (here 'Pekin 1860', 'Taku Forts 1860' and 'Canton 1857').

Americans and Russians, all of whom also had trading agreements to protect. The resultant treaty of Tientsin had negligible effect and when a deputation went to China the following year it was forcibly rebuffed. The British and French governments thereupon resolved upon a strong joint expedition under British command to settle the matter by advancing on Peking itself along the route of the Peiho River.

The force amounted to 10,000 British and 8,000 French. The British contingent of two divisions comprised Queen's units from both England and India, including three Royal Engineer companies (8th, 10th and 23rd), together with Indian army units. Sapper interest centres on the 2nd Division led by Major General Sir Robert Napier, fresh from his triumphs in the Mutiny, and on the Taku forts (see page 71). These had already been attacked twice in the war, only once successfully. The initial landings were made some distance away and two other fortified towns had to be overcome en route. The sappers were kept busy keeping the army moving, bridging rivers and canals, and supporting the assaults. The taking of the Taku forts themselves called for ingenuity and heroism in equal measure, but after a four-hour bombardment and gallant assault the Chinese surrendered. They then brought calamity on themselves by imprisoning the party sent forward to negotiate the final terms, and executing two of them. After two more fierce battles Peking itself was occupied, and revenge for the treatment of the envoys was taken by the destruction of the Summer Palace. In the final post-war settlement Kowloon was added to the ceded area of Hong Kong. The 99-year lease of the New Territories was agreed in 1898, just before the Boxer Rebellion.

The British garrison that then remained in China cooperated with the Manchu rulers in putting down the Taiping rebellion that had ravaged the country for some sixteen years. Captain Charles Gordon (q.v.) was at that time engaged in surveys in areas affected by the rebellion. He so distinguished himself in the course of these, and the hazardous engineer reconnaissance that he undertook, that he was offered command of the 'Ever Victorious Army', the irregular Chinese force raised to fight the rebellion. His extraordinary display of energy and courage in the subsequent operations was instrumental in

SPOILS OF WAR

The Summer Palace in Peking (Beijing) was destroyed in 1860 by the Anglo-French expeditionary force. The then Captain Charles Gordon was present and recorded his experiences in one of his letters:

Owing to the ill-treatment of the prisoners experienced at the Summer Palace the General ordered it to be destroyed, and stuck up proclamations to say why it was ordered. We accordingly went out, and after pillaging it burned the whole place, destroying in a Vandal-like manner most valuable property which could not be replaced for four millions … It was wretchedly demoralising work for an army … Everybody was wild for plunder. You would scarcely conceive the magnificence of this residence, or the tremendous devastation the French have committed. The throne and room were lined with ebony carved in a marvellous way; there were huge mirrors of all shapes and kinds, clocks, watches, musical boxes with puppets on them, magnificent china of every description, heaps and heaps of silks of all colours, embroidery and as much splendour and civilisation as you would see at Windsor … The French have smashed everything in the most wanton manner. It was the scene of utter destruction which passes my description.

Right: A 'lotus flower' bowl cut from a single piece of jade; roof tile in the form of a Qilin, a mythical beast that served to ward off the danger of fire; and a deity of long life made to celebrate the 70th birthday of the emperor Chien Lung.

FIELD MARSHAL LORD NAPIER OF MAGDALA, GCB, GCSI (1810–1890) shown in field marshal's uniform. Robert Napier joined the Bengal Engineers in 1826 and thereafter most of his service was in India. Much of his early career was spent in construction engineering in the challenging conditions that India offered. This experience was put to good effect in the Sikh Wars (see page 53). But his flair for field command emerged in the fighting in the aftermath of the 1857 mutiny (see page 61) where, as a brigadier general, he conducted a series of brilliant campaigns that snuffed out resistance in Central India. Almost immediately he took command of the 2nd Division in the China War of 1857–60 and the task of taking the Taku forts (q.v.) fell to him. After five years as military member in the Council of the Governor-General, Napier was appointed Commander-in-Chief of the Bombay army. It was from that post that he was selected to command the expedition to Magdala that brought him such renown at home and his peerage. He was Commander-in-Chief India from 1870 to 1876 and after leaving India returned home and served as Governor of Gibraltar until 1882. He became a field marshal in 1883.

bringing the revolt to an end in the area for which he was responsible. This episode led to great public acclaim and the career that was to end so dramatically on the steps of the Governor's palace at Khartoum (see page 83).

The Boxer Rebellion (1899–1901) was an uprising by a movement of disaffected Chinese who called themselves 'Fists of Patriotic Union' and directed their hostility against all foreigners. Action to stem a rising tide of outrages was put together piecemeal by the nations affected, by far the strongest forces coming initially from the Japanese and Russians. The British were once again otherwise engaged, this time in South Africa, but eventually were able to field more than 18,000 men in an international force that amounted to some 68,000. The prelimi-

Left: Leopard-skin regalia, a cope or 'lembd', placed on Sir Robert Napier by the Kasa, John of Tigree, on conclusion of the treaty at the end of the Abyssinian campaign in 1868.

Left: 10 Field Company was the only unit sent from Britain to join the Abyssinian Expedition of 1867–8. On their strength were the first official army photographers (sappers) to join an overseas expedition. The photograph shows some of the special equipment making its first appearance in war such as signal lamps and telegraph items.

nary British land operations were conducted by the sapper Colonel Arthur Dorward, with much distinction, until the main force under General Gaselee arrived. Three companies of Sappers and Miners from India provided the bulk of the engineer contribution. The usual combat tasks kept the sappers busy during the operations which saw Tientsin besieged and, eventually, Peking occupied. A major challenge then had to be faced to provide for the garrison that remained in an exceptionally cold winter. Among other notable achievements was the renovation of the Tientsin–Peking railway, another international effort in which the British share was fourteen miles of reconstruction, 29 miles of new construction and 11,500 feet of bridging repaired or laid anew.

ABYSSINIA

In January 1864 King Theodore of Abyssinia, who had established himself on the throne by force twelve years earlier, took the British consul in Abyssinia captive, along with several other Europeans. He had been offended at the British government's lack of response to his overtures to Queen Victoria seeking help in recovering the holy places in Palestine. Months of delicate negotiations for the release of the captives came to nought. By early 1867 they had been taken to Magdala, Theodore's fortress 10,000 feet in the mountains and 400 miles from the coast. Britain decided to force the issue and appointed Lieutenant General Sir Robert Napier to command an expedition to rescue the prisoners.

Below: Zula, the small port on the Red Sea coast, selected by Lieutenant General Sir Robert Napier as his base for the Abyssinian Expedition of 1867–8. Some 14,000 troops with their logistic support and engineer stores passed through this port. A 60-mile road and 10-mile railway were built forward of the port.

Above: The Abyssinian War Medal 1868 is an oddity among campaign medals, being thinner and smaller than others. It was one of the most expensive medals produced because of the delicate work of stamping the name of the recipient on the reverse. The medal illustrated was awarded to Lieutenant T. J. Willans.

Above: Storming the Gateway, a watercolour depicting the assault on Magdala. The artist, Major Frank James of the 20th Native Infantry (Indian Army), was Deputy Assistant Quartermaster General on Sir Robert Napier's staff.

Right: The Magdala Vase, a silver-gilt vase made from a silver drum taken at Magdala on 13 April 1868. It was presented by Lord Napier to the Mess of the Queen Victoria's Own Madras Sappers and Miners as a memento of the service in Abyssinia of the detachment consisting of G, H and K Companies under the command of Major H. N. D. Prendergast VC.

Right: The Tibet Chorten, a replica of the Tibetan religious emblem found throughout the country. It commemorates the 1903–4 Younghusband mission to Lhasa for which the sapper Major General James Macdonald was the commander of the 3,000-strong military 'escort', which included seventeen officers of the Corps and a company each of the Bengal and Madras Sappers and Miners.

The most serious enemy was likely to be the forbidding terrain so Napier's 14,000-strong army contained thirty Royal Engineer officers and seven companies of Indian Sappers (three Madras and four Bombay) together with 10 Field Company, Royal Engineers. They built a port, a railway and roads. They reconnoitred the route, signalled their findings by flag, lamp and telegraph, made maps, took photographs and supplied water by condensation at the base and from tube wells on their route. Pre-fabricated water tanks, produced in conjunction with the Royal Navy, anticipated the Braithwaite tank of later years. The army had to fend off just one Abyssinian attack before the final assault. This took place on 13 April 1868 after a brief bombardment. It was led by sappers carrying the necessary tools, ladders and bags of powder to break through the defences. Theodore, too proud to go into captivity, took his own life, the prisoners were unharmed and Robert Napier returned to a hero's welcome and a peerage. The campaign has passed into history as an 'Engineers' War'.

THE TIBET MISSION (1903–4)

In what was effectively another 'Engineers' War', Colonel Francis Younghusband (King's Dragoon Guards), the explorer and intelligence officer, led the diplomatic mission to Lhasa that was intended to persuade the Tibetans to discuss the increasing influence of Russia in their affairs. The sapper Brigadier General J. R. L. Macdonald,* who had been responsible for the railway work in China, was appointed to command the strong force that would convince the reluctant Tibetans to negotiate. The Tibetans resisted but could not hold back Macdonald's small army (1,000-strong but well-armed). The bloodshed on both sides caused some controversy at home. But nobody could deny the extraordinary achievement of the sappers in seeing the mission with all its impedimenta (10,000 coolies and as many draught animals) to its objective across four mountain passes 14,000–17,000 feet in height, and several rivers including the mighty Tsang-po, while also supporting the operations to break through the enemy defences.

*In 1891, when still only a captain, Macdonald had earned his spurs in East Africa.

13 | ASHANTI AND ZULU

The nineteenth century growth of the British Empire inevitably led to boundary disputes. In two cases in Black Africa these escalated to full-scale wars against native armies with a powerful military tradition reinforced by autocratic discipline. Both were fought in difficult unfamiliar terrain where engineering laid the foundations of success.

THE ASHANTI WAR (1873–74)

The Ashanti people occupied the forest hinterland of the Gold Coast (modern Ghana). By the middle of the nineteenth century they had encroached on to the coastal plain and established trading rights with the European stations there. Using their powerful army they then sought to set up their own bases in what was a British protectorate by treaty with the local chieftains. The British government responded to this challenge by dispatching a small force under Colonel Sir Garnet Wolseley to raise local levies, free the Protectorate from Ashanti control and then take the war into their territory to destroy their capital at Kumasi.

This would require a 100-mile march through thick forest with little more than a single-file track to follow. Amongst the special equipment taken by the sappers to support this enterprise were Blanshard pontoons (see page 43), telegraph equipment, prefabricated bridges and a steam sapper, the first traction engine to go on active service. In the event they had to depend much on improvisation and local materials with which to bridge the multitude of water gaps and construct fortified camps along the route. With the sappers leading the route-clearance parties, in the course of which Lieutenant Mark Bell won the

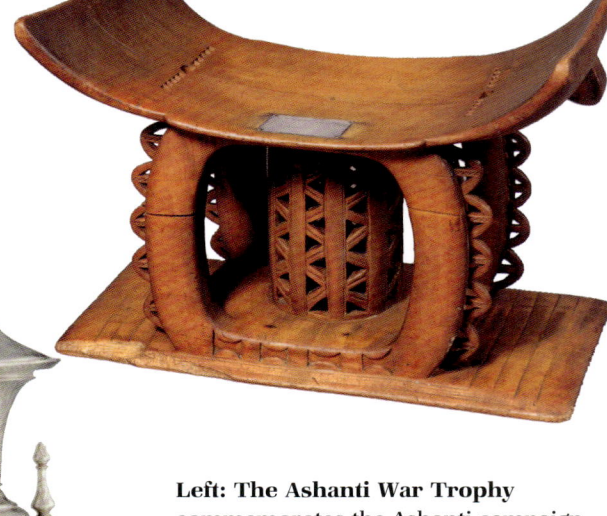

Left: The 'Steam Sapper' (see page 107) first saw active service in the Ashanti War. This model is of Aveling and Porter Steam Sapper No. 3, which was delivered to the School of Military Engineering in March 1872.

Left: Ashanti Stool, taken during the First Ashanti War (1873). The tribal stool in Ashanti symbolised the power of the chief. That of the Asantahene, king of the nation, was known as the Golden Stool and its capture became the objective of several (unsuccessful in this respect) expeditions to assert British supremacy in the Gold Coast.

Left: The Ashanti War Trophy commemorates the Ashanti campaign of 1873, which led to opening up the hinterland of what is now Ghana. The 28th Company RE, which was the only engineer unit of the British force, cut a road through virgin jungle from the Coast to Kumasi and was thus in the vanguard throughout the advance.

Right: The Ashanti Star 1895–6 was earned by more than sixty members of the Corps who served in that war. (See page 80.)

VC (see page 68), Wolseley reached Kumasi in early February 1874. He eventually secured a treaty with the king (the Asantahene) under which he would not take up arms against Britain and 'use his best endeavours to check the practice of human sacrifice.'

Above: The Ashantee Medal 1874 was issued to some 90 members of the Corps who served in Wolseley's expedition (see text). The reverse is noteworthy in that it has more intertwined living and dead bodies, some fighting either side of a tree, than on any other medal reverse. One clasp 'Coomassie' was issued to those present at the battle of Amoaful (about half the Royal Engineers).

THE PRAH RIVER

The Commanding Royal Engineer was Major Robert Home. His report on the problem of crossing the river states:

'The bridge now became the great question. The Blanchards [sic] pontoons were the only means available for crossing, and were far too few to rely on for the vast number of carriers that had to pass to the front. Twelve trussed girders had been prepared at Chatham. These were very slight and strong, and would give a bridge across the river 2 ft. 6 in. wide; but considering the very important link in the communication this bridge was, it was determined to make the bridge not less than 5 ft. wide. Thus the trussed girders would give one half of the roadway required, and the remainder, consequently, would have to be made up of material to be procured on the spot. Captain Buckle had made a section which showed that the river was not only deep, but that the bottom was fairly regular, and of hard sand. There were no means of driving piles, and such light trestles as the tackle on the spot would get out would undoubtedly be carried away in a short time, as the current was running nearly four miles an hour, and freshets were to be expected. Under these circumstances, crib-work piers were determined on.

'The day after the first crib was launched it was found to have sunk 17 inches into the mud, and to be only 4 inches out of level. This was easily made up by putting a couple of thick sticks on the low side. The crib was raised to its full height, and transoms of roughly adzed timber laid in.'

Top: Prefabricated spans being tested at Chatham before the expedition. They can be seen in the first three spans from the right in the photograph, right.

Above: Blanshard infantry pontoons. These were smaller than the engineer version (see page 43)

Right and below: The Prah river was the principal water obstacle on the route to Kumasi. Top, a photograph from the 1873 expedition showing the prefabricated spans and cribs; bottom, a watercolour by Lieutenant Colonel H. M. Sinclair from the 1900 expedition showing the use of cask pontoons.

THE SECOND AND THIRD ASHANTI WARS AND THEIR AFTERMATH

A second expedition was mounted in 1895 in which the new, and less compliant, Asantahene was forced to accept British protection. He was exiled to the Seychelles. However, the colonial government felt that their authority over the Ashanti people was incomplete without acquiring the Golden Stool, in which the soul of the Ashanti nation was believed to reside. Their vain attempts to capture this prize led to the Third War in 1900. The resultant animosity was somewhat moderated by the successful tour as governor of the former sapper Sir Gordon Guggisberg (q.v.) during which the foundations of the future Ghana's prosperity were laid.

THE ZULU WAR (1879)

A brave, disciplined, and well-trained army combined with a certain scorn for human life were also features of the Zulu nation that challenged British plans for ordering affairs in southern Africa in 1879. Boundary disputes and border troubles convinced the British government that peace and prosperity in the region could only be achieved after the subjugation of the Zulus, led by their chief, Cetshwayo. To that end an army invaded Zululand in three columns in January 1879 under the overall command of Major General Lord Chelmsford, who also led the central column.

The early disaster of Isandlwana (*see* Durnford portrait below), in which the central column's base camp was overrun with the loss of over 1,300 lives, and the redeeming defence of Rorke's Drift (q.v.) are well-known. In the war as a whole the sappers, comprising four field companies (2nd, 5th, 7th and C Troop 30th) were fully stretched when the columns were able to continue their advance. Notable among the engineering achievements was the crossing of the Tugela river to support the southernmost column and the building of the defensive position at Eshowe in which this force was beleaguered for some months. A similar work at Khambula secured the northern column against a potentially overwhelming attack on the scale of Isandlwana.

Lord Wolseley, with earlier experience as Governor of Natal, was sent out to rescue what was thought to be a disastrous situation. But the fighting was effectively over, after the Battle of Ulundi, by the time he arrived.

Above: The sword and waterbottle used by Lieutenant John Chard at Rorke's Drift.

COLONEL ANTHONY DURNFORD

Colonel Anthony Durnford, a Royal Engineer, commanded the 540-strong column of the 1st Battalion Natal Native Contingent, a rocket battery and three companies of Natal Native Horse which had been ordered to join Chelmsford's central column on its initial advance into Zululand. He advanced from the Isandlwana base camp and discovered an impi of 20,000 Zulus preparing to attack. Durnford and all his men were killed in an heroic fighting withdrawal. When the remains of the fallen were recovered four months later, the order to Durnford's column was recovered from his body and is now in the RE Museum.

Right: Anthony Durnford's South Africa Medal 1877–79. The reverse of this medal depicts a lion, representing Africa crouching in submission in front of a Protea bush. Above are the words 'South Africa', and in the exergue there is a Zulu shield and four assegais. Seven dated clasps were issued.

Far right: The orders found on Durnford's body, scraps of which have survived, go some way to understanding why he decided to move forward to face the Zulu impi forward of Lord Chelmsford's base camp at Isandlwana. Illustrated is the original order for his column to move to Rorke's Drift in support of No 3 Column.

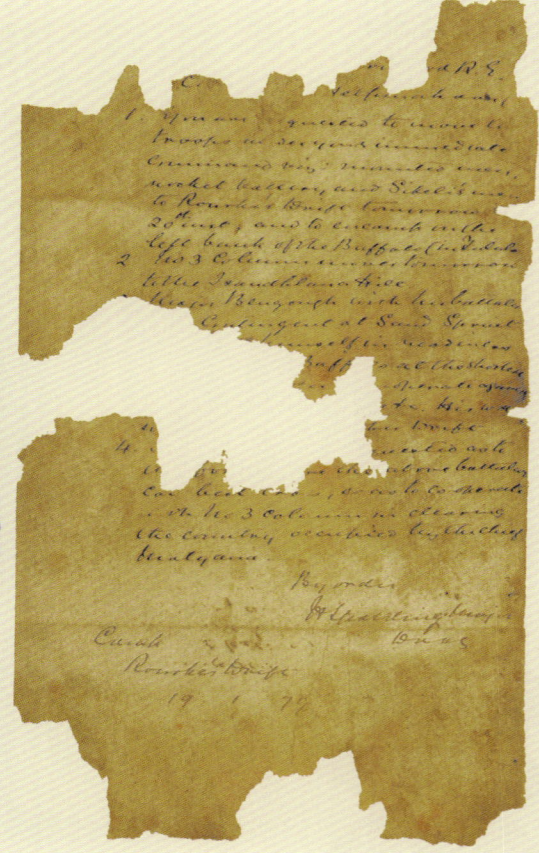

Above: The Zulu War centrepiece, completed in 1881 to commemorate the Corps' role in the war. One of the three models of the Zulu warriors is a likeness of Ohan, brother of Chief Cetshwayo.

Above: Brigadier General Sir Gordon Guggisberg, KCMG, DSO (1869-1930) was appointed Governor of the Gold Coast in 1919. He knew the area well from survey expeditions earlier in his career. Taking advantage of flourishing cocoa production and sales, he oversaw the development of the infrastructure of the country and set it on the road to eventual independence in 1957.

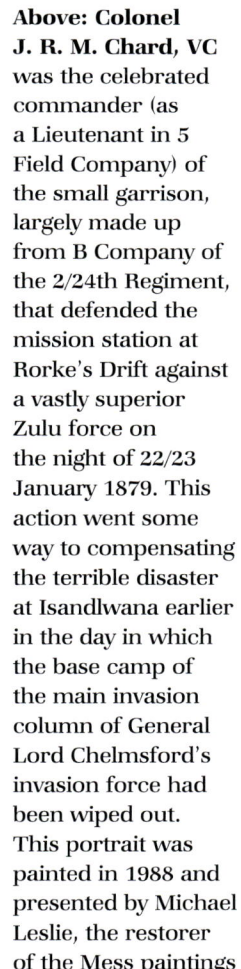

Above: Colonel J. R. M. Chard, VC was the celebrated commander (as a Lieutenant in 5 Field Company) of the small garrison, largely made up from B Company of the 2/24th Regiment, that defended the mission station at Rorke's Drift against a vastly superior Zulu force on the night of 22/23 January 1879. This action went some way to compensating the terrible disaster at Isandlwana earlier in the day in which the base camp of the main invasion column of General Lord Chelmsford's invasion force had been wiped out. This portrait was painted in 1988 and presented by Michael Leslie, the restorer of the Mess paintings after the 1975 fire.

Above: Rorke's Drift mission station, as sketched by Lieutenant Chard for the report he was invited to write for Queen Victoria. Chard was aware of the impending threat of attack by a Zulu impi and realised that there was little choice but to stand and fight from the best defensive position he could create. He ordered the construction of the famous wall of mealie bags and some upturned wagons to connect the hospital building (13) and storehouse (8). The first attack came from the hill (Oscarberg, 1) against the hospital. When this had to be abandoned after the main attack from the garden (15), the perimeter was reduced to the area of the storehouse by another wall built from biscuit boxes. In the end 4,000 Zulus were held off by the garrison of 140 of whom thirty were hospital patients and only eighty-one were a formed body of soldiers (B Company of the 2/24th). Eleven VCs were awarded for the action, including Chard's.

14 | CONFLICT ON THE NILE

The dust had barely settled on the Second Afghan and Zulu Wars when Egypt and the Sudan began to take centre stage in the theatre of British strategic interest, and with them two of the leading players, the Royal Engineers Charles Gordon and Herbert Kitchener. The route to India was the heart of the matter. In 1875 Britain had acquired Egypt's shares in the Suez Canal and by 1882 sought, with France, to bring order to the chaotic rule of the Khedive. The anti-western revolt that followed under Arabi Pasha was defeated in a brief war that ended at the battle of Tel el Kebir. Wanting to be part of the action, Kitchener, who was then a just-promoted captain engaged on the survey of Cyprus, claimed some sick-leave, went to Egypt and took part in a short clandestine operation. Shortly after his return to Cyprus a request arrived for his transfer to the Egyptian Army as second-in-command of a cavalry regiment. Thus, thanks to his own subterfuge, began his spectacular link with the region that was to end as Commander-in-Chief (Sirdar) of the Army, victor of the battle of Omdurman; and then, through command in South Africa, bringing the Boer War to its conclusion and Commander-in-Chief India to his final renown as Secretary of State for War in 1914 until his death in 1916.

When Kitchener first went to Egypt in 1882, down in the Sudan insurrection was being fomented by the Mahdi whose 'Dervish' (now, more correctly termed Ansar) army destroyed a 10,000-strong British-officered Egyptian force in the desert west of Khartoum in September 1883. Farther east towards the Red Sea another fanatic, Osman Digna, was rising in support. The British government had no wish to become committed in the Sudan, but had to become concerned for the safety of the Egyptian garrisons there.

In response, General Sir Charles Gordon, seventeen years Kitchener's senior, was sent to Khartoum in 1884 with instructions to evacuate the garrisons but was besieged before being able to take any action. At the same time, a force was sent to the Red Sea area, under the command of Major General Sir Gerald Graham VC, where operations were conducted against Osman Digna. Public concern at Gordon's fate resulted in the belated despatch of General Sir Garnet Wolseley to mount the relief expedition, in which the 34-year-old Kitchener displayed outstanding bravery as an intelligence officer, spending much of his time in disguise spying ahead of the main body. The spearhead of Wolseley's force, under the command of the sapper Colonel Sir Charles Wilson (a colleague of Kitchener's from

GENERAL CHARLES GORDON spent many long hours on the back of a camel during the six years he spent as Governor of the southern Sudan and Governor-General in Khartoum (1874–80). He covered prodigious distances, particularly in his efforts to halt the slave trade. This statue, the work of Onslow Ford, ARA, is solid bronze and weighs fifteen tons. It was erected in front of the present Headquarters RSME and unveiled by the Prince of Wales on 19 May 1890. It carries the following inscription:

Charles George Gordon, Royal Engineers, Companion of the Bath, Major-General of the British Army, Mandarin of China, Pasha of Turkey, Governor-General of the Soudan. Born at Woolwich, 28th January, 1833; killed at Khartoum, 26th January, 1885. Erected by the Corps of Royal Engineers, 1890.

A replica statue was originally set up in St Martin's Place, London, in 1902, but moved to Khartoum later that year. It was returned to England with the Kitchener statue in 1958 and taken to Gordon School, Woking.

Above: Gordon's Pasha's Coat.

THE LAST DESPATCH

portrays the imagined scene on 15 December 1884 when General Charles Gordon hands over his journal and final letters to the captain of the paddle-steamer *Bordein* on the last journey permitted by the Mahdi, whose forces had already besieged Khartoum. Forty days later he was killed by the Mahdi's troops.

Left: The famous letter to his friend Colonel Charles Watson which begins:

My dear Watson, I think the game is up, and send Mrs Watson, you and Graham my adieux. We may expect a catastrophe in the town on or after ten days' time …

GORDON'S LAST STAND as depicted by the artist G. W. Joy. The exact circumstances of Gordon's death are not known but this version is no less plausible than any other. Beside this is General Gordon's Khartoum Star, an example of the decorations that Gordon designed and had struck within the besieged city of Khartoum for gallantry in his defending army. In the centre is a grenade surrounded by an inscription in Turkish reading 'The Siege of Khartoum'. It was made in silver-gilt, silver and pewter by a local goldsmith Bishara Abdel Molak. (LMG 38880, Leeds Museums and Galleries)

his Palestine days), just failed to reach Khartoum before Gordon was killed in the Governor-General's palace. Thus arose a national outcry and demands for the revenge that it was to be Kitchener's destiny to satisfy.

For twelve years the Sudan was left in the hands of the Mahdi while the Egyptian government rebuilt its economy and its army. The latter owed much to the energies of Kitchener, who became its Sirdar in 1892. While the British government still wanted nothing to do with the Sudan, in the 'scramble for Africa' the French and Italians both had their eyes on it. After a preliminary move in 1896, during which Dongola province was re-occupied, Kitchener moved south by measured steps until the stage was set for the final showdown with the Mahdists at Omdurman on 2 September 1898.

SAPPERS IN SUPPORT

Five Royal Engineer field companies had been included in Wolseley's force for the 1882 war, together with a pontoon troop, a telegraph

Left: 26th Field Company embarked on the Nile during the 1884 Gordon Relief Expedition. Wolseley decided, without engineer advice, to use the Nile route to Khartoum in preference to the alternative desert crossing from the Red Sea littoral. Eight hundred whalers designed to carry twelve men with rations for 100 days were ordered from boat-builders in England. The firms received their order on 12 August and the last boats left England on 3 October. At Alexandria they were handed over to Thomas Cook & Son who had been awarded a contract for taking them the 800 miles into the Sudan by rail, barge and steamer tow. At their destination (Saras) they were prepared by the sappers and rigged by the Royal Navy for their onward journey carrying the troops. (Original print by Melton Prior, RE Museum)

Right: The Egypt Statuette, commonly known as Cleopatra, but representing an unknown queen of the Ptolemaic dynasty, commemorates Sir Garnet Wolseley's campaign of 1882 against Arabi Pasha.

troop, a field park and 8th (Railway) Company. One field company was allotted to each division (24th and 26th), the remainder (17th, 18th and 21st) acting as Corps Troops. An Indian contingent also provided two field companies of Madras sappers. The railway company was an innovation, the first such unit in the Corps. The men were trained by arrangement with two British railway companies before embarking for Egypt. The telegraph section was something of a novelty too. The telegraph that had been available in both Abyssinia and Ashanti had technical limitations. In Egypt it materially contributed to the speed of operations. Tel el Kebir was also the first occasion on which a commander-in-chief's victory report was sent by telegraph direct from the field. The message to Her Majesty, the Secretary of State and others was sent at 8.30 a.m. on 13 September; the reply was received at 9.15 a.m.

The strength of Wolseley's 1884 Nile Expeditionary Force was limited by logistics to about 9,000. Two field companies (11th and 26th) were allocated, together with 8th (Railway) Company and a section of the Telegraph Battalion. While in this operation the railway played little part in the advance south, it was valuable during the subsequent withdrawal.

Graham's force on the Red Sea in March 1884 had just one field company, but when he returned the next year to lead an expedition to Khartoum from the Red Sea, he was well provided with sappers. As well as the 17th and

Right: Sir Herbert Kitchener in 1890, a portrait by Sir Hubert von Herkomer and Frederick Goodall showing him as a colonel in Royal Engineer khaki drill uniform shortly after his victory over the Mahdi's forces when he was Governor of Suakin.

24th Field Companies RE and one from the Madras Sappers there were two telegraph sections, the 10th Railway Company and a balloon section. 24th Field Company distinguished themselves in the battle of Tofrek, a furious fight early in the campaign when a detachment under Major General McNeill was surprised by the Mahdist forces. The balloon section, very much experimental in its first operational appearance, proved its worth in the captive observation role. A local narrow-gauge railway system was successfully built, but a civil contract for a standard-gauge link from Suakin to Berber only achieved sixteen miles before the expedition was called off for political reasons.

In contrast to this and to 1884, the railway that Kitchener ordered and brought about to supply and reinforce the army for the Omdurman campaign was a strategic masterstroke. As

Above: Lieutenant General Sir Gerald Graham, VC, GCB, GCMG (1831–1899) was one of the first recipients of the VC, awarded for his selfless gallantry in the maelstrom of the failed storming of the Redan in the Crimea in June 1855. He also saw active service in the 1857 Mutiny in India, in China (1860) and Egypt (1882) where he commanded a brigade. Later he took command of the forces on the Red Sea littoral, based at Suakin, and conducted successful campaigns against Osman Digna, the fanatical but elusive follower of the Mahdi. He received the thanks of Parliament, was made GCMG and retired from the army in 1890.

Above: A Mahdist robe, taken at the battle of Atbara (19 April 1898), Kitchener's victorious prelude to Omdurman. The followers of the Mahdi wore garments with simple stylised patches signifying poverty.

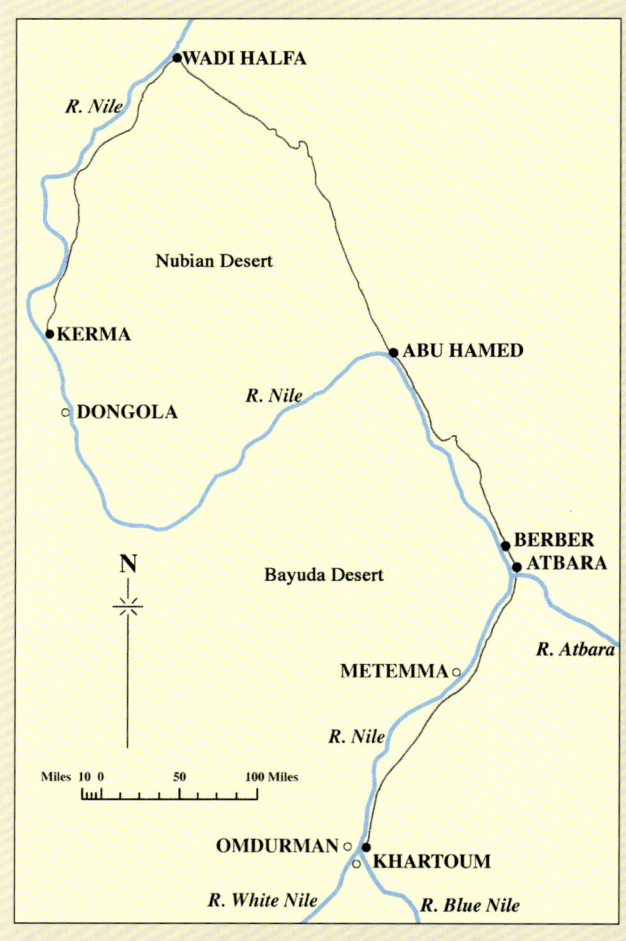

Colonel Sir Percy Girouard KCMG DSO, a railway engineer, was recruited as a lieutenant by Kitchener to lead the team building the Nile railway that guaranteed victory in the Sudan War. Girouard was a French Canadian with a transatlantic lack of deference for authority. Kitchener liked his direct manner, infectious enthusiasm and sense of humour, qualities that were to motivate the other members of the team. A life-long bond of understanding grew up between them to their mutual benefit in later conflicts.

The Sudan Desert Railway (576 miles) was built under the supervision of Royal Engineers between January 1897 and December 1899. This included a four-month break around the battle of Omdurman. The section to Kerma had been built the previous year to support Kitchener's Dongola expedition and some of the materials were taken up for the main line across the Nubian desert. An essential adjunct to the railway was the telegraph, which generally followed the same route.

Above: The Egypt Medal 1882–89 covered the actions at The Nile in 1884 and 1885. The reverse depicts the Sphinx on a pedestal with the word 'Egypt' above. The first medals issued had in the exergue the date 1882. For the second campaign, 1884–90, the same medal was used but undated.

Above: The Khedive's Star was awarded to all recipients of the Egypt Medal. This five-pointed bronze star depicts the Sphinx with the three Great Pyramids in the background. There were four different issues of the star, three with the dates 1882, 1884, 1884–6; and one undated issue.

Winston Churchill put it in *The River War*: 'On the day that the first troop train steamed into the fortified camp at the confluence of the Nile and Atbara rivers, the doom of the dervishes was sealed.' Two years earlier Kitchener had begun his advance south with the 1896 Dongola expedition. For this he created a railway battalion. A number of young Royal Engineer officers were appointed to key posts to build and manage the railway that eventually developed into the famous desert railway that cut across a loop of the Nile to support the 1899 campaign.

Apart from the railway team and the all-important telegraph sappers that accompanied the advance south, only one field company (2nd) joined Kitchener's 23,000-man force for the final battle of the campaign at Omdurman in September 1898.

15 | BOERS AT BAY

The South African or 'Great Boer' War (1899–1902) arose from the British government's determination to assert its authority over the Transvaal, which it had never regained since its rebuff at Majuba Hill in the First Boer War (1880–1). The Boer leaders, however, sought total independence in their now gold-rich country. A row developed over the treatment of foreign nationals in the Transvaal, which ended in an unacceptable Boer ultimatum requiring all recent military reinforcements to leave South Africa within 48 hours. The Boers advanced into Natal and battle was first joined at Talana on 20 October 1899.

Ladysmith was besieged as the commander in Natal, Sir George White, tried to check the Boer advance. A major front then opened up as

Above: The Kitchener statue outside Kitchener Barracks, Chatham, is a replica of one produced for India by Sydney March. During Kitchener's visit to the Sudan in 1911 the Governor-General, Sir Reginald Wingate, suggested that Khartoum should have a statue of him. When Kitchener objected that the cost was not justified, Wingate offered several tons of brass from the cartridge cases collected from the field of Omdurman to which tin could be added. This was accepted and the statue cast free of charge. It weighs five and a half tons (hollow-cast) and is one and half times life size. Delayed by the First World War, it was erected in 1921. It was returned to Great Britain in 1958 and unveiled on 25 April 1960 by the Secretary of State for War, the Right Honourable Christopher Soames, CBE, MP.

Lieutenant General Sir Redvers Buller's force tried to break through to its relief. Buller had been appointed to the command of an army corps that had been assembled in the UK to deal with the emergency. On his arrival at the Cape, he had decided to take personal command on the Natal front with the bulk of the army, leaving operations in the Free State and Cape Colony in the hands of his subordinates, Lieutenant Generals Lord Methuen and Sir William Gatacre respectively.

At the same time the Boers also put Mafeking and Kimberley under siege. A series of early setbacks for the British army included a single 'Black Week' in December. In this Methuen, aiming at the relief of Kimberley, was defeated at Magersfontein; Gatacre, defending the north-eastern border of Cape Colony, lost much of his force in an ill-judged manoeuvre; and Buller was badly defeated at Colenso in a premature attempt to break through to Ladysmith. Reinforcements were sent out and Field Marshal Lord Roberts was appointed to the overall command. He was already aged 67 and to provide him with a younger deputy, the now Major General Lord Kitchener (aged 49), fresh from his triumphs in the Sudan, was appointed Chief of Staff.

Roberts's strategy was first to relieve Kimberley and support Buller in his efforts in Natal, then advance on Pretoria along the axis of the main railway line through Bloemfontein by means of rapid outflanking movements by the Cavalry Division. The Kimberley operation succeeded in February 1900 and was followed by the trapping of a 5,000-strong Boer force under Piet Cronje at Paardeberg. During this time some of the most dramatic events of the war were being enacted on the Natal front. Boer attempts to break the defensive ring round Ladysmith were frustrated by aggressive action, particularly at Wagon Hill. There the exploits of Lieutenant Robert Digby Jones, commanding a detachment from 23 Field Company setting up a gun position, earned him a posthumous VC. At Spion Kop the newly arrived 5th Division under the sapper Lieutenant General Sir Charles Warren came agonisingly close to a break-through. Ladysmith was eventually relieved on 28 February 1900, the day after Cronje's surrender at Paardeberg. Two weeks later, on 13 March, Roberts entered Bloemfontein on the heels of the Boers. That very morning the railway to the north had been blown up in a dashing raid under Brevet Major Aylmer Hunter-Weston, commanding the 1st Field Troop of the Cavalry Division.

In early May Roberts launched the advance that took him to Pretoria. The Boers evacuated

Above: The Bethulie railway bridge was destroyed by the Boers in March 1900 during General Gatacre's advance towards the Orange Free State. The nearby road bridge had also been prepared for demolition. Captain Grant with Lieutenant Popham of the Derbyshire Regiment removed the charges in the face of the enemy. The road bridge was used for trains for some time before the rail diversion shown in this photograph, critical for the support of Roberts's advance, could be completed.

Above: Tugela pontoon bridges. In General Sir Redvers Buller's attempts to relieve the siege of Ladysmith from December 1899 to February 1900, several pontoon bridges were built across the Tugela River by A Troop of the Bridging Battalion using the Clauson or Mark II bipartite pontoon, in service from 1889 until the First World War. The painting by Georges Scott depicts the final successful crossing in February.

Right: Bird's Eye view of Norval's Pont, a watercolour by an unknown artist illustrating two bridges over the Orange River. The Boers seized the railway bridge in November 1899 and blew it up. The pontoon bridge was built by C Troop of the Bridging Battalion the following March. The railway bridge was open by May 1900.

the city and Roberts entered on 5 June. By this time the Boer leadership had already decided to adopt the guerrilla tactics that were to tie so many troops down protecting the lines of communication, and that were to extend the war for a further two years. By September 1900, however, the Transvaal had been occupied and annexed to the British Crown and little appeared to be left other than operations to mop up the remaining Boers at loose around the country. Roberts and Buller returned to Britain; Kitchener took over as Commander-in-Chief.

Kitchener now sought to defeat the guerrilla 'commandos' by separating them from the

THE EVERARD SKETCHES
Little is known about Lance Corporal Everard apart from his skill as a draughtsman, as these sketches show, one sent home from the Boer War, the other illustrating a field company training in England.

Mounted Sappers, blowing up a Bridge.

Mounted Sappers, travelling 'Light'.

Above: Blockhouses were first introduced in January 1901 to protect the railways from attack by the Boers. Sapper-designed, the first ones were erected by contract. The octagonal design illustrated, designed by Major Spring Rice, commanding 23 Field Company and introduced in February 1901, was developed later into a circular form. Kitchener early appreciated the value of lines of blockhouses connected by wire fences as a means of controlling the movement of the guerrilla units. The first cross-country line, connecting the railways, was planned in June 1901. By the end of the war thirty such lines had been completed with fences of ever-increasing complexity coupled with a variety of ingenious alarm systems and rifles and machine-guns on fixed lines.

civilian population on which they depended for succour, the approach that was to be adopted in so many counter-insurgency campaigns in the second half of the century. The railway and telegraph system provided the framework for this, backed by the network of blockhouses that burgeoned to impede enemy movement (see page 89). Mobile columns now became the main manoeuvre formations replacing the divisional organisation that had served well for the conventional phase of the war. Thus was Boer resistance eventually worn down until Kitchener negotiated peace terms in May 1902 that launched the new South Africa as part of the British Empire.

SAPPER SUPPORT

Sapper support for these operations was more diverse than in any war hitherto. The eleven infantry divisions that eventually took part each had a field company, the Cavalry Division only a field troop. Back-up was provided by Army Corps troops in units with such specialisations as bridging, telegraph, survey, ballooning, field

park, railways, photo-reconnaissance, search-lights and steam transport – the electricians even brought out the first mobile X-ray unit. The scale of the commitment may be judged by the fact that over 8,000 men of the Corps qualified for the Queen's South Africa medal.

In addition to their usual field engineering tasks the divisional units had their fair share of fighting as infantry, as Sir Ernest Swinton (then a captain) evoked so colourfully in his *Defence of Duffer's Drift*. At Wepener near the Basutoland border, for example, Major Cedric Maxwell with a few sappers found himself in command of a force of 1,800 troops when they were attacked by the famous Christiaan De Wet with about 8,000 of his guerrillas. Skilfully designed field defences and the maintenance of communications enabled the small garrison to hold out until reinforcements came and De Wet withdrew.

That the railways would become the life-blood of the army was appreciated early and Major Percy Girouard of Sudan fame (see page 86) was appointed Director of Railways for the South

Above: A Newman and Guardia 5×4 twin-lens camera with telescopic lens, used by the world's first operational military photo-reconnaissance unit, commanded in South Africa by Lieutenant Charles Foulkes.

Right: Ballooning (see also page 106). The Boer War marks the zenith of oper-ational ballooning. Three sections were formed and deployed between September 1899 and March 1900. They proved an asset, mostly for artillery observation, until the guerrilla phase of the war, when they were broken up. 1st Section was allocated to the Kimberley relief operations and then marched to Pretoria through Bloemfontein. It was in action until August 1900. 2nd Section was at Ladysmith and performed valuable service for 27 days until its gas ran out. 3rd Section operated on a variety of tasks until May 1900. Illustrated is a balloon from 2nd Section oper-ating at Ladysmith.

Above: An armoured traction engine and howitzer. Most of the traction engines sent to South Africa initially were shipwrecked on the way out. Only a limited road transport service, made up from local sources, was possible until the losses could be made good later in the war. Several of the traction engines, which were operated by 45 Field Company, were fitted with armour to protect the driver from snipers. The armour of this model has been removed to show the detail of the Fowler B5 with its towed 6-inch howitzer.

Below: Armoured trains. From the start of the war it became clear that trains must be armed and, where possible, armoured, to protect them from Boer raids. That illustrated is No. 17 Armoured Train, mounting one quick-firing 12-pounder and two 0.303 Maxims.

African Field Force. His department controlled the railways in British-held territory, arranged for the repair of lines damaged by the enemy and took over the lines in Boer territory as they were captured. The organisation for all this embraced Royal Engineer units (two railway companies and a number of fortress companies allocated for this task), elements of the existing civilian railway company organisations and a special Railway Pioneer Regiment raised in the country. Major J. E. Capper commanded this regiment with Captain Ernest Swinton as his adjutant, two

Above: The South Africa Bowl. This fine piece comprises a large open bowl around which is a frieze of a veldt wagon drawn by a team of eighteen oxen. The statuettes on the plinth represent two sappers and two Boer figures. The latter were modelled from the life-size statues from South Africa that for some years adorned the front of the Institute building but were later returned and now form part of the Kruger Memorial in Pretoria.

Above: The South African War Memorial Arch was unveiled by King Edward VII on 26 July 1905. It carries the names of 420 officers and men of the Corps who died in the war. The sculptured panels in high relief, by W. S. Frith, depict a blockhouse, a movable ox-wagon blockhouse, a pontoon bridge with a balloon in the background and a bridge diversion.

Right: The Queen's and King's South Africa Medals, 1899–1902 and 1901–1902 respectively. All soldiers, including over 8,900 sappers, who participated in the Boer War of 1899–1902 qualified for the Queen's South Africa Medal. Those who served there after the death of Queen Victoria also earned the King's Medal. The production of these medals caused great controversy because they were initially issued with the dates 1899–1900 on them. The modifications were made at huge expense when it was clear that the war would continue longer than expected. The Queen's death caused further argument when the new King insisted on a special medal to be struck with his head on it. The reverse of the King's Medal is the same as the later versions of the Queen's.

officers who later distinguished themselves in other fields and rose to high rank (see pages 108 and 117).

The telegraph was likewise critical to the success of operations. Similar to the railways, a specialist telegraph 'division' was set up and arrangements made for coordination with the civilian organisation. The lines were an easy target for Boer raids, and their maintenance and repair involved much hard and dangerous work by the repair teams, often operating in enemy-held territory.

Ballooning, for all its practical problems, proved the value of aerial reconnaissance. Steam traction made a limited contribution and experience was gained in this field. Searchlights were successfully used for communication during the Ladysmith siege by reflecting flashes from the clouds.

An unusual feature of this war was the 'double-hatting' of the Chief Engineer, Brigadier-General Elliott Wood, as commander of the lines of communication. Initially he had about 11,000 men distributed along the railway line for this purpose. In December 1899, in this role, he had to eject an incursion of Boers from Cape Colony; later he established a fortified post across the Orange River, the first such occupation of the enemy's territory. Wood then moved forward as Roberts's Chief Engineer, handing over command of the lines of communication to his successor, Brigadier-General Henry Settle.

Left: Field Marshal the Earl Kitchener of Khartoum and Broome, KG, KP, GCB, OM, GCSI, GCMG, GCIE, PC, as a major general in 1900 after his victory at Omdurman and before setting off to South Africa as Chief of Staff to Lord Roberts. His involvement in the Middle East began with surveys in Palestine and Cyprus during which he became a competent Arabic speaker. His rapid rise in the Egyptian Army (see page 82) came to an end in 1899 when he was called to join Roberts in South Africa, himself taking over as Commander-in-Chief in November 1900. His resolute pursuit of the Boer guerrilla teams finally brought peace. Four years as Commander-in-Chief in India then followed, distinguished by his adamant stand against the Viceroy, Lord Curzon's, attitude to army reforms. In 1909 he left India as a field marshal and in 1911 took up the appointment of Consul General in Egypt. On the outbreak of the First World War he became Secretary of State for War. His vision and inspiration in raising army manpower and initiating the expansion of arms production had a marked effect on the war before he was drowned while setting out on a mission to Russia in June 1916 after HMS *Hampshire* struck a German mine off Scapa Flow. (Portrait by Sir A. S. Cope, PRA)

III
PLOUGHSHARES
TO SWORDS

16 | CREATIVITY

The Corps' ability go to war in 1899 (see page 87) with the range of skills and equipment described in Chapters 18 to 20 had its roots nearly a century earlier in the days after the defeat of Napoleon. Then quite suddenly, the British Army had few operational commitments and while the infantry and cavalry were obliged to spend their days mainly on drill and duties, the Royal Engineers were able to offer a pool of trained talent available for the service of the nation. They represented a substantial proportion of the country's skilled engineers. Despite dropping from a strength of 262 to 193, this was about the same number as the membership of the Institution of Civil Engineers at the time. It was not difficult for government departments to accept the employment of such a scarce commodity. Likewise the Royal Sappers and Miners; among the arguments for their creation had been the assertion that '… the Corps can never become a burden on this Country – as when there are no armies in the Field the men can be employed on works going forward in England …'

PRISONS AND HOSPITALS

In the 1830s there was a pressing need for prison reform and plenty of theories on which to base this. In 1837 Captain Joshua Jebb was called in as architectural adviser to the Inspectorate of Prisons and to design Pentonville according to new principles. One of these was the total separation of prisoners from most human contact other than the prison staff in the early part of their sentence. Individual cells were therefore needed with heating, ventilation and water closets. Jebb's revolutionary design for these services, which have attracted the admiration of modern experts, became standard in all prisons built during the rest of the century, as did his layouts devised to ensure full supervision. Jebb was soon invited to become a commissioner of Pentonville. After a decision to separate military from civil prisons he also became Inspector-General of Military

Above, left: Lieutenant General Sir William Denison, KCB was the quintessential sapper of the post-Napoleonic era, in which the British Army offered little opportunity for action in the field. Commissioned in 1826, he was a product of Pasley's forcing house, complete with its course on architecture (see also page 44). On the one hand, he was a brilliant engineer and experimenter in new techniques. He was awarded the Institution of Civil Engineers' Telford prize for his experimental work on Canadian timber while working on the Rideau Canal (see page 101); as Superintendent of the Admiralty Works Department at Portsmouth, he was also a pioneer in testing new wrought-iron designs. On the other hand, he earned renown as a colonial governor, first in Van Diemen's Land (he was still only a captain but had been knighted for his work for the Admiralty) and later in New South Wales and Madras.

Above, right: Major General Sir Joshua Jebb, KCB had been seconded to the Home Office in 1837 as a captain, so founding a tradition of sapper involvement in the design and management of the prisons.

Prisons. In 1850 a centralised Board of the Directors of Convict Prisons was formed and Jebb became its chairman.

This post remained in sapper hands for the rest of the century, Jebb's successor being another sapper, Lieutenant Colonel Edmund Henderson. He had just returned from Western Australia where he had spent thirteen years as Comptroller of Convicts in a new colony forming up towards the end of the policy of transportation. Henderson's command there, initially solely for the control of convicts, eventually expanded into the much wider development role of infrastruc-

94

Above: Captain (later General Sir) Edmund Du Cane (1830–1903).

Above right: Worm-wood Scrubs, completed in 1883 under the super-vision of General Sir Edmund Du Cane, symbolises the achievements of sapper involve-ment in the prison service that began with Captain Joshua Jebb's model prison at Pentonville in 1842. Wormwood Scrubs was built entirely by convict labour based in the old Millbank prison, which it replaced.

ture building, for which a company of sappers was sent out including many married men who were to settle in the colony. One of the company officers was Lieutenant Edmund Du Cane who was destined to become the last sapper Surveyor General of Prisons. Captain Du Cane in fact took over from Henderson in 1869 when the latter was appointed Chief Commissioner of Metro-politan Police, a post he held with immense success until eventually handing over to another sapper, Colonel Charles Warren (later Lieu-tenant General Sir Charles).

Less surprisingly, sappers were also drawn into the development of barracks and mili-tary hospitals on the wave of improvements demanded by the Army Sanitary Commission that operated after the Crimean War. Captain Douglas Galton was appointed to this and served on it until his death in 1899, loaded with distinc-tions including a knighthood. He resigned his

commission in 1862 to become Assistant Perma-nent Under Secretary of War, an appointment that may have been influenced by Florence Night-ingale (whose cousin was married to Galton) in her drive to bring about sanitary reform. Galton is best known for his innovative approach to heating and ventilation that revolutionised barrack standards and was incorporated in the design of the new Herbert Hospital at Woolwich, in which he played a leading part.

THE ADMIRALTY WORKS DEPARTMENT

The Admiralty Works Department was another natural habitat for talented sappers. They brought to it an enthusiasm for innovation, a practical approach to planning and manage-ment and a rigorous testing regime. Their involvement covered the period from 1837 when Captain Henry Brandreth was appointed to the post of Director of the Department, to the late 1860s. Brandreth had had some experience of working with civilian iron companies over a project to introduce cast-iron framed buildings into the West Indies in the 1820s. The use of iron, and sapper enthusiasm for it, was a particular feature of this period; so much so that when a Royal Commission on the Application of Iron to Railway Structures sat following a serious bridge collapse, the Royal Engineers on the commission were the only members qualified to advise.*

A notable application of iron was in the covered slipways of the dockyards. Wooden-cov-ered slip sheds had been in use since 1812, but iron offered significant improvements in both durability and potential span. Colonel Godfrey Greene, a retired Bengal Engineer, who was one of Brandreth's successors in the Admiralty Works Department, introduced the practice of in-house

*Brandreth himself, who died suddenly during the proceedings, Douglas Galton, John Lintorn Simmons (the future field marshal), Henry James and Henry Harness, all later knighted for their various services.

Left: Covered Slipways at Chatham, now part of the Historic Dockyard, demonstrate the progress of the design of wide-span roofs fostered by the Royal Engineer representation in the Admiralty Works Department. The right hand one is No. 3, an early wooden roof. Next come Nos 4, 5 and 6, the product of Major Brandreth's time as Director. No. 7, a substantial advance, was the work of Colonel Godfrey Greene.

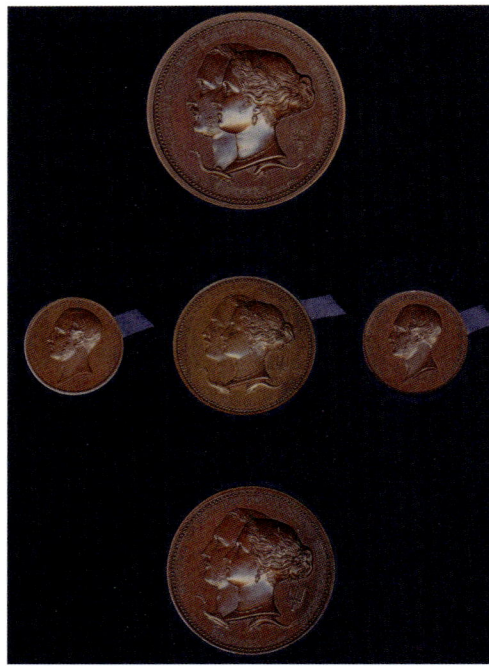

Above: The Great Exhibition of 1851 was marked by a medal. That awarded to Lieutenant Henry Tyler for his services to the committee is illustrated. Shortly after the Exhibition he was appointed an Inspector of Railways and held the post for twenty-four years both inspecting and advising on railways at home and abroad, being knighted for his services in 1877.

Above, above right and right: The South Kensington Exhibition Hall was adapted from an original design by Francis Fowke as a permanent home for future exhibitions. Erected quickly using pioneering techniques, it was only ever used for the 1862 International Exhibition and afterwards demolished (by sappers using explosives) to make way for the Natural History and Science Museums. These photographs are the work of Sergeant Mack, one of the senior ranks who learned their remarkable skills in the Science and Art Department.

Left: Captain Francis Fowke (1823–1865) was the sapper architect who made the most significant mark on the profession in the 19th century. His early career showed a flair for inventive ingenuity. As a result, in 1855 he was appointed as an Inspector to the Science and Art Department, which had been formed in the aftermath of the Great Exhibition. He played a major part in the early development of the South Kensington Museum (the present Victoria and Albert). He had many design commissions and was specially admired for his creative use of cast and wrought iron and glass. Just before his death at only 42 years of age, he had won the competition for the design of the new Natural History Museum and the distinguished civilian architect Alfred Waterhouse's scheme was adopted instead.

The Fowke Medal was introduced by the Institution in memory of Francis Fowke in 1866, to be awarded annually to a young officer considered to have specially distinguished himself in the School of Construction. Now reproduced in bronze, it is awarded to the top student in each of the Clerk of Works (Construction), (Electrical) and (Mechanical) and Military Plant Foreman's courses.

Above: The Lewis Medal, originated in 1918 in memory of Colonel J. F. Lewis, was awarded to the young officer of each batch 'for proficiency in fortification and field engineering' at the School of Military Engineering. This dedication is inscribed on the reverse of its unusual rectangular design, along with a representation of a fortification in Gibraltar. The medal was discontinued in 1932.

design rather than by contract. Not only did he bring wide-span design to a peak of excellence, but his experience in working with the leading contractors of the day led to what is generally accepted as the first multi-storey iron building to adopt the portal frame principle. The Boat Store at Sheerness was completed in 1860 and has stood there ever since, modestly unaware that it incorporated what was to become universal 20th-century practice.

THE 1851 GREAT EXHIBITION AND ITS AFTERMATH

The 1851 Great Exhibition provided another opportunity to exploit sapper talent. Lieutenant Colonel William Reid, then Commanding Royal Engineer at Woolwich, was appointed Chairman of the Executive Committee.* Later, twelve more officers joined him in various capacities along with two companies of Sappers and Miners. In the aftermath of the Exhibition, the Science and Art Department was established under the Board of Trade, one of whose objects was to create an architectural 'atelier' as a national centre for promoting and coordinating design. Central to this was to be a Museum of Construction to which Captain Francis Fowke (q.v.) was appointed Engineer and Director. Thus began the South Kensington Museum, better known now as the Victoria and Albert.

*Reid had been a subaltern in the Peninsular War where he and a fellow-officer, Peter Wright, had impressed Wellington who referred to them as 'my favourites "Read" and "Write"'. Since those days he had filled several colonial governorship appointments but had returned to regimental duty. Sir William Reid KCB died in 1858.

While Henry Scott (q.v.) took over Fowke's baton in 1863, sappers were much involved in a range of activities around the country, much of it embracing new technology. Innovations were made in heating and ventilation that were incorporated by Jebb into Pentonville and its successor prisons from the 1830s to the end of the century. Another pioneer in this field was Douglas Galton (see page 95), who had designed a revolutionary grate for barrack heating and who, after 22 years' commissioned service, retired as a captain to continue in his career as a civil servant, rising to become Director of Public Works and Buildings. Much of this work arose from the army sanitary reform movement following the Crimean War.

ARCHITECTURE AT HOME AND ABROAD

All these elements converged in the field of architecture, the skill which the Duke of Wellington as Master General of the Ordnance had decreed in 1825 was to be taught at Chatham. This initiative, embraced so dynamically by Pasley, took root and flourished round the world. Sappers grasped the opportunity to solve their problems by the use of new materials, as in the introduction of pre-fabricated iron-framed buildings in tropical areas (see page 99). The elements of architecture were to hand in the pocket books of every sapper officer who, far from home, could tackle any task from a field latrine to a Palladian mansion. India, with its insatiable demand for fine public buildings, barracks and churches, gave great scope for creative sappers and many made a lasting mark elsewhere throughout the Empire.

Professionally, however, fortifications was the most serious business for corps architectural expertise in the second half of the 19th century. This was influenced by the resurgence of the threat from France, by the arrival of steam propulsion in hostile navies and advances in artillery and the development of applications for new materials such as iron and concrete. (See page 110.)

Right: Major General H. Y. D. Scott (1822–83) took over much of Charles Pasley's experimental work on cements at Chatham, during which time he discovered and patented a new form of the material that was produced commercially. In 1864 he succeeded Francis Fowke (q.v.) in the Science and Art Department but as 'Director of Works' rather than as 'Architect and Engineer', Fowke's title when he died. Scott's most celebrated achievement was the Royal Albert Hall, which had been sketched out by Fowke before his death. The final design for this can be credited to Scott's genius for bringing together acknowledged experts in their field and coordinating their work into a successful result. Scott retired from the Army in 1871, but continued with the Science and Art Department until his job was cut in 1883 for which he received no recompense. He died the same year.

Left: The Royal Albert Hall, completed in 1871, was principally the work of Major General Henry Scott from original sketches by Captain Francis Fowke.

Left: The Frieze depicting various scenes connected with the arts is made from terracotta mosaic tiles. The artists' designs were photographed by Sergeant Spackman, a sapper photographer working in the Science and Art Department, and the images, enlarged to about six feet, were the basis for laying the mosaic by the ladies of the South Kensington Museum mosaic class.

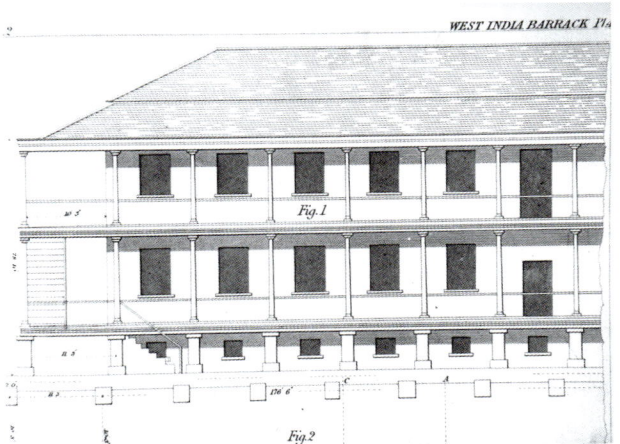

Left: The Silver Mint and St Paul's Cathedral, Calcutta were both designed by Captain (later Major General) William Forbes of the Bengal Engineers. In their contrasting Classical Grecian Doric and Gothic styles they exemplify both the skill and versatility of sapper architectural accomplishments throughout India; they also reflect the 'battle of the styles' that characterised the profession at home in the 19th century. The construction of the Silver Mint was held up by Forbes being called to duty at the 1826 siege of Bhurtpore (see page 53), a further demonstration of versatility. It took six years to build and to install the complex equipment.

Above left, and left: Victoria Barracks, Sydney was built c.1840 by the Colonial Engineer for New South Wales, Lieutenant Colonel George Barney. The framework is cast iron of the same design that had been specially developed in the 1820s for the West Indies where Barney had worked under Major General Sir Frederick Smith.

Right: Sir John Burgoyne in informal mode, probably in South Kensington, a portrait by the distinguished photographer James Fenton. Presumably they met again in the grimmer circumstances of the Crimea.

Above: The Channel Tunnel boring machine designed by Captain Thomas English in collaboration with Colonel Frederick Beaumont. It started cutting a seven-foot-diameter bore on 16 October 1880. The project was joined by two more machines in 1881, but in 1882 the scheme was called off by order of the Board of Trade. About 2,950 yards had been achieved.

17 | INFRASTRUCTURE

The Corps bequeathed to the world an astonishing legacy of roads, railways and canals in the course of the 19th and early 20th centuries. Chapter 7 of Volume III of the Corps' History illustrates the scale of the commitment by listing more than ninety posts held by sappers on detachment to the Foreign and Colonial Offices, excluding India, from the 1880s until the First World War. The list includes fifteen governor or lieutenant governorships and several 'commissioner' or 'consul-general' posts. Many more were for 'special duties', usually for a technical advisory role, and for attachment to the Public Works Department. Many of these works, however, date from as early as the 1830s.

There were two main incentives: the urgently needed development of the fledgling colonies, and the often more pressing strategic and internal security demands of the countries concerned.

CANADA AND AUSTRALIA

In Canada, for example, ever since the War of 1812 (see page 29) it was clear that the British settlements on the northern shore of Lake Ontario could easily be isolated if the United States were to cut the vital St Lawrence lifeline. Security could best be guaranteed by a waterway joining the Ottawa river with Kingston using the navigable portion of the Rideau River and certain lakes that lay on the

route, the rest to be joined by canals. A project for this was begun in 1826 and took five years to complete (see opposite).

By contrast, the development of the settlement of British Columbia was a deliberate act of colonisation following the discovery of gold on the Fraser River. The first sapper party of an officer and twenty men arrived some four months ahead of the 1858 Boundary Commission (see page 67). By May the following year the strength of what became known as 'The

Right: Building the Cariboo Road, a romanticised portrayal by the Canadian artist Rex Woods of one of the toughest jobs undertaken by the Columbia Detachment of the Royal Engineers (see text). Gold had been found in the Cariboo mountains in the headwaters of the Fraser River whose valley proved the most favourable route for a road to reach the workings. The Detachment had carried out several road projects earlier, but this was on a different scale altogether involving prodigious amounts of blasting and the building of long sections supported on cribs.

Left: Lieutenant Colonel John By (1779–1836) and the Rideau Canal. John By was a 46-year-old lieutenant colonel on half pay, planning a comfortable retirement in the country, when he was ordered to Canada to undertake the construction of the Rideau Canal from the Ottawa River to Kingston (see text). He had spent eight years in the country before, had operational experience in the Peninsula at the siege of Badajoz, and had to his credit several fine engineering achievements including the extensive new works for the Royal Armoury at Enfield near Waltham Abbey.

The Rideau project began in 1826. By's operating base grew into a small township that became known as Bytown and developed eventually into Canada's capital city, Ottawa. His force comprised eleven other Royal Engineers, two of whom commanded the two 81-strong companies of Royal Sappers and Miners raised specially for the project. Labourers and contractors' men made up the total employed at any one time to some two thousand. In five and a half years they constructed 52 dams, 47 locks and eighteen miles of newly dug canal along the 123 miles of waterway, at a cost of £800,000.

By's astonishing success in welding together this disparate force, many of whom succumbed to disease from the unhealthy environment and in overcoming the engineering difficulties of a hostile terrain and climate, was ecstatically acclaimed in Canada at the end of the project. But on his return, instead of receiving the promotion and knighthood that he deserved, By was unjustly criticised by the government for over-expenditure and mismanagement. He died a wronged and disappointed man aged only 56 years.

Right: General Sir Alexander Taylor, GCB (1826–1912) was commissioned from Addiscombe in 1843. He spent most of his service in India. During his early career in India he saw plenty of action and gained a reputation for amazing feats of courage and physical strength. Among his many accomplishments was the extension of the Great Trunk Road of some 300 miles through undeveloped country. At the siege of Delhi in 1857, as a captain and second-in-command to Baird Smith, he was responsible for planning the storming of the Kashmir Gate (see also page 60). Much of the success of the operation can be attributed to him. At the time Baird Smith was still suffering from the effects of a severe wound. In 1865, Taylor was appointed Chief Engineer in the Punjab. In 1876 he became Deputy Inspector General of Military Works and president of the Defence Committee. After retirement he became president of the Royal Indian Engineering College, Cooper's Hill, the college set up in 1872 largely to train civilian engineers. He did much to improve the college during the sixteen years he held the appointment.

Columbia Detachment', under Colonel Richard Moody,* had risen to five officers and 150 men. They remained in the colony until 1863 when the detachment was disbanded. Each soldier and officer was offered his discharge in the country plus a grant of 500 acres. Although all the officers returned home, most of the non-commissioned officers and all but fifteen of the sappers stayed behind to make a new life. Their most celebrated achievement was the Cariboo Road (q.v.) but their presence, though expensive to the colony, was a force for stability and made a significant contribution to the early life of British Columbia. In Australia colonisation arose around the convict settlements. Individual Royal Engineers went out, first to Van Diemen's Land in 1835 and shortly afterwards to New South Wales. Two distinguished governors in these early days were Sir George Gipps (New South Wales 1838–46), a Peninsular veteran, and Sir William Denison (Van Diemen's Land 1847–55 and New South Wales 1855–60). Western Australia lacked the benefit of a Royal Engineer until 1850 when Captain Henderson arrived in the Swan River colony with five Sappers and Miners, followed in due time by 20th Company Royal Engineers, one of whose officers was Second Lieutenant Du Cane (see also page 95). Their main tasks were

*In an earlier appointment Moody had been Governor of the Falkland Islands, giving his name to Moody's Brook.

roads and bridges and eventually public buildings, many of which still survive.

INDIA

India, with all its pressing needs, set the pattern for this work. The Royal Engineers as such were not involved until the post-Mutiny incorporation of the Bengal, Madras and Bombay Engineers into the Corps. Before that time, young men destined for those corps were trained at Addiscombe, the Company's military academy, but also attended the same special-to-arm course at Chatham as their Royal Engineer contemporaries. Involve-

Above: The Lindsay Cup symbolises the debt that the infrastructure of India owes to the civil works of military engineers. It is a very fine cup of Indian silver, seventeen inches high. The lid is surmounted by a figure of a Hindu god, while two similar figures on each side of the bowl serve as handles. The bowl has embossed decoration representing scenes and figures from Hindu mythology. An unusual and beautiful feature is the Serpentine baluster of perforated silver around the rim. The whole piece is of exceptionally fine workmanship and a remarkable example of the art of the Madras silversmiths. The following inscription is engraved on the lid: 'Presented to Colonel J. G. Lindsay, RE, Agent and Chief Engineer Southern Mahratta Railway on his retirement from the service, by the officers whose names are inscribed on the pedestal, as a token of their esteem and regard. April 1901.'

Below: The mules of the exploring expedition – Sappers' travelling mess, a scene of the British Columbia detachment painted by Sir Howard Elphinstone, VC. Elphinstone was appointed tutor to HRH Prince Arthur, Queen Victoria's youngest son and the future Duke of Connaught, whom he accompanied on an official tour of Canada.

Above: Major General Sir James Browne, KCSI, CB (1839–96) was commissioned into the Bengal Engineers in 1857. He took part in several frontier campaigns and worked on the Punjab section of the Grand Trunk Road. Most of his career was spent on the North-West Frontier on a variety of engineering projects. There he acquired such an affinity with the tribesmen that he was able to move freely in the area and his intelligence reports found favour in government circles. He took part in the Second Afghan War as a political officer and in 1882 went to Egypt as Commanding Royal Engineer of the Indian contingent in the Tel el Kebir campaign. In 1884, Browne was promoted brigadier general and put in charge of the construction of the Sind–Pishin railway (q.v.) (see text). Browne was Quartermaster General of the Army in India from 1889 to 1892. His final appointment was as Agent-General and Chief Commissioner in Baluchistan. He devoted his characteristic energy and humanity to effecting material improvements to that country, but his health failed and he died in harness at the age of 56.

Above: The Louise Margaret Bridge (a contemporary sketch with a photograph of the site taken in 2003). The bridge crosses the Chappar Gorge (or 'Rift') on the Harnai section of the Sind–Pishin railway. Two hundred and forty feet high, it had nine spans, one of 150 feet. The engineering and administrative difficulties of this project were stupendous. In all the Harnai line was 224 miles long and had, in one section, to rise 6,000 feet in 120 miles (double the rise of the St Gotthard railway) and traverse the three-mile gorge passing through nine tunnels. Cholera and malaria ravaged the 20,000-strong work force and at one point the four companies of sappers were so reduced that they were unable to mount guard. The railway was completed in 1887 and in March of that year the bridge was formally opened by HRH The Duchess of Connaught, after whom it was named. She was accompanied by the Duke, who had started his military career in the Royal Engineers.

ment in 'works' was a professional necessity and it has already been shown how compatible this could be with engineering in the field (chapters 8 and 16).

The Great Trunk Road is the outstanding case of the concurrence of strategic necessity and economic benefit. In the 1830s roads were a novelty in India, but the then Governor General of Bengal, Lord William Bentinck, ordered one to be built from Calcutta to Delhi to improve communications. Its strategic importance, allowing the rapid movement of troops, soon became obvious and in due course its route was established all the way from Calcutta to Peshawar (over 1,500 miles) and beyond, eventually reaching Kabul. It was built in sections during most of the century, but enough was complete to prove its value during the 1857 Uprising when the siege train and reinforcements were able to reach Delhi from the Punjab

in sufficient time to effect the siege. Many distinguished sappers had contributed to this work in their earlier careers, including Lord Napier, Sir Alexander Taylor and Sir James (Buster) Browne.

Water transport was superior to road for most of the 19th century. Moreover, canals brought prosperity through irrigation. This art engaged the efforts of hundreds of sappers and at least one distinguished gunner, Sir Proby Cautley, who built the Ganges Canal. The foremost sapper exponent was Sir Arthur Cotton (q.v.) whose truly remarkable achievements made sufficient local impact for a statue to be erected in his memory in 1983, a rare and possibly unique honour for an Englishman since Independence. With his professional eye Cotton had also observed some weaknesses in the design of the Ganges Canal and he was able to incorporate improvements in his own work on the Cauvery, Godavari and Krishna river systems. When Colonel Sir Colin

Right: The Russian Punch Bowl was presented to Colonel Sir Colin Scott-Moncrieff (q.v.) for his services to the Russian government in 1890 in advising them on irrigation in the region of the city of Merv (now know as Mary, Turkmenistan). This exquisite piece is of consummate craftsmanship and probably has no equal in Britain. It was made of silver-gilt cloisonné by Pavol Ovchinnikov in Moscow by Imperial commission in 1833. The gift came as a total surprise to Sir Colin when it was delivered to him at Brindisi, en route to Egypt, by M. de Ranner, the Czar's representative. It was presented to the Mess by the Scott-Moncrieff family in 1928.

Left: Krishna Annicut from the quarry Bezoarah looking South, a sketch of one of Sir Arthur Cotton's irrigation projects on the Krishna River.

Right: Sir Arthur Cotton (1803–99) was one of the leading irrigation engineers of his day whose work, particularly in the United Provinces, now the state of Andhra Pradesh, brought about such immense economic benefit to the region that his name remains honoured there to this day. He held an almost evangelical belief in the importance of waterways, not only for irrigation but also for transport, for which he argued they provided a more economic solution than railways, in which he also had experience. Cotton, a close relative of Lord Combermere (see page 53) earned his operational spurs in the First Burma War (see page 70). His first major irrigation scheme was in present-day Tamilnadu, but by the late 1840s he had launched the Godavari project in Andhra Pradesh and from that was appointed Chief Engineer of Madras Presidency. He retired in 1877 and enjoyed nearly forty years more of life and the opportunity vigorously to promote his ideas from England.

Scott-Moncrieff, as Chief Engineer of the Ganges Canal, undertook modifications to Cautley's great work he had already established a world-wide reputation as an irrigation engineer. This led to his appointment, after retiring from the Bengal Engineers, to rebuild the great Nile barrage and its surrounding canals. This in turn led to the invitation from the Russian government to visit central Asia and advise on the irrigation of Merv.

Railway engineering in India followed naturally from experience on the roads and also occupied many sappers, often as consultants or managing directors in the early days. Men like Lieutenant General Sir Richard Strachey were highly influential in pushing forward railway development. Work on the ground was not for the faint-hearted, involving as it did fighting with unforgiving terrain, disease and the movement and control of huge labour forces. One of the sapper-built railways needed almost exclusively for military reasons was the Sind–Pishin line connecting the Indus valley with the strategic plateau that separates Quetta from Kandahar. Forty years later the Khyber Railway (q.v.) was built for much the same strategic reasons (see chapter 10).

UBIQUE

These few examples stand out to illustrate the phenomenon of the sapper's mark left around the globe from exploits in peace and war that continue to the present day. The Bailey bridge (see page 163) is probably the most ubiquitous symbol of this endeavour, but more often than not today's traveller will use roads and railways unaware of the significance of their origins.

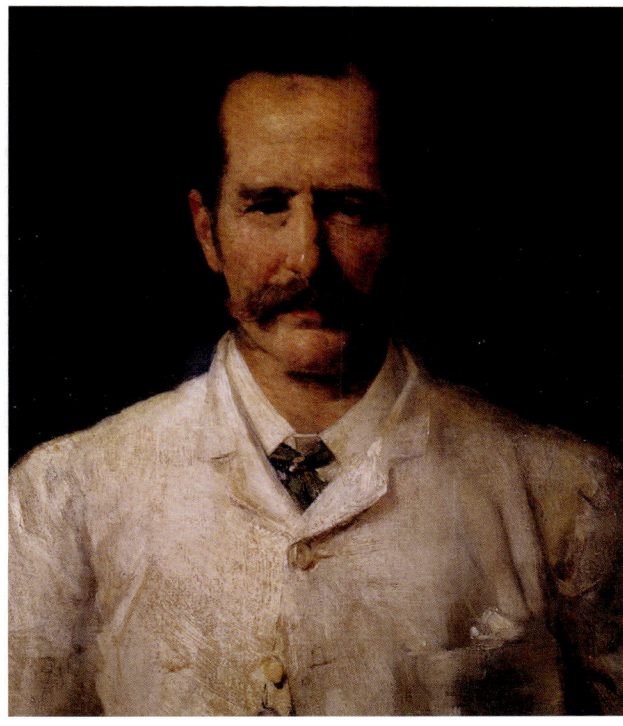

Above: Colonel Sir Colin Scott-Moncrieff, KCMG, KCSI (1836–1916) was commissioned into the Bengal Engineers in 1856 and, after five hectic months of active service during the Uprising, joined the Public Works Department. He specialised in irrigation engineering and during his first leave undertook a tour of France, Spain and Italy from which he published a book, *Irrigation in Southern Europe*, 1868. He became Executive Engineer of the eastern Jumna Canal and in 1869 Superintending Engineer of the Ganges Canal. He was appointed Chief Engineer, Burma in 1881, and retired in 1882. While on leave in Egypt on his way home from India he was offered and accepted the post of Under Secretary of State for Public Works. In the process of restoring the Nile barrage and the irrigation works of Lower Egypt, he also succeeded in abolishing the corvée forced labour system. He retired from Egyptian service in 1892, was appointed Under Secretary for Scotland 1892–1902 and finally Chairman of the Indian Irrigation Commission.

Below: A section of the Khyber Railway under construction. It was built in the aftermath of the Third Afghan War to provide for reliable and swift resupply for any force that might in future years have to be moved through the Khyber valley in response to any hostile move by the Afghans or the tribes in the border area. Lieutenant Colonel Gordon Hearne was in charge of the project. When complete it was twenty-seven miles long with two and a half miles of tunnels and seven crossings of the Khyber River.

On 13 August 1914 four squadrons (44 aircraft) of the two-year-old Royal Flying Corps began crossing the Channel to the Western Front. Behind them lay 33 years of enterprise and far-sighted endeavour by the Royal Engineers, and the inspired individuals who worked under their auspices, culminating in the formation of the Air Battalion, Royal Engineers in 1911.

It all started with balloons in 1878. Ballooning was already a popular pastime in civil life for the well-to-do and authority was now given for the experimental production of the equipment that might be needed for military use in the field. The task was given to Captain Henry Lee. The matter had been studied before this. For example, Captain Beaumont (see page 99) had been sent across to America during the Civil War to report on practice in the Federal Army. One energetic mastermind, who joined the original team and stayed in the business almost throughout, was James Templer, then a captain in the Middlesex Militia. He was also much engaged with the development of steam traction (see box).

In 1891 the now established Balloon Section and Depot had been formed and set up at Aldershot. Most of the practical difficulties of manufacturing balloons and producing hydrogen in the field had been overcome. Two experimental deployments overseas had been accomplished in 1885; detachments under Majors Henry Elsdale

Left: Lieutenant Colonel J. L. B. Templer, KRRC, an infantry militia officer, was the moving spirit behind much of the development of both steam traction and aeronautics in the Corps. An enthusiastic amateur balloonist who served with the Royal Engineer balloon units throughout the existence of that branch of the Corps, he turned to steam traction to provide the mobility for the equipment, mostly hydrogen gas cylinders, that was needed to support balloon sections in the field. Thus he also became an expert in this matter and it was in that guise that he served in the Boer War. After the war Templer returned to aeronautics, resuming his role as, effectively, a civilian consultant to the Balloon Factory until retiring in 1908. Although the 'father of ballooning', he received no official recognition or reward for his achievements. He died, aged 78, on 2 January 1924.

Left and above: Ballooning became a Royal Engineer activity from its experimental days in the 1880s through to the formation of the Royal Flying Corps in 1912. The photographs show a balloon ascent in 1893 and the practicalities of filling a balloon at Aldershot in 1903. The manufacturing of hydrogen in the field (normally by electrolysis) and transporting it (under pressure in cylinders carried on 'tube wagons') was a major factor in the military development of ballooning.

Above: The Weinling family were the only people in Britain who understood the technique of joining the sheets of gold-beater's skin, keeping it a closely-guarded secret. This material derived from the blind gut (caecum) of the ox and had properties that made it ideal for the envelopes of balloons and airships. The Weinlings, Alsatian immigrants, who earned their living making toy balloons, were persuaded to move from the East End of London to Chatham where they worked for over thirty years in government employ.

and Templer went respectively with the Bechuanaland expedition and the Sudan Expeditionary Force. As it happened, both force commanders were sappers: Major General Sir Charles Warren and Lieutenant General Sir Gerald Graham.

Optimistic reports from these led to a full deployment in the South African War. Three sections were sent out and proved their value in artillery observation. The 1st Section operated with the Kimberley column and greatly unsettled the Boers at both Magersfontein and Paardeberg. The 2nd Section was caught in Ladysmith and was effective until it ran out of gas. The 3rd Section had a brief moment of glory before being

STEAM SAPPER

The Corps pioneered steam traction for military purposes. By itself it was never a war-winner but it paved the way for mechanical road transport and was an integral part of the practical development of military ballooning. The first Corps engine, christened Prince Arthur in honour of the Queen's youngest son who joined the Corps in the same year, was acquired in 1868. All the early tractors, known as 'steam sappers' until 1894, were supplied by Aveling and Porter of Rochester. The first engine was required to power machinery but also to haul a five-ton load up a slope of 1 in 12. As well as the obvious sapper applications there was clearly potential for these machines for hauling heavy artillery. The second machine was bought for this purpose, and christened Steam Sapper No. 2.

The first to go to war was No. 8, despatched to the Gold Coast with Wolseley's force in 1873 (see illustration). Its performance was disappointing but the need of a support train for military ballooning accelerated further developments. At the same time highly successful trials had taken place in India led by the 'Superintendent of the Government Steam Train', the dynamic and inventive Colonel R. E. B. Crompton (page 108).

The Boer War (pages 87–92) represented the zenith of steam road transport in war. Colonel James Templer was appointed Director of Steam Road Transport. A new Royal Engineer company, the 45th, was established specifically for this purpose.

By the end of the war almost fifty engines were in service. However, a War Office Mechanical Transport Committee had been set up and the decision was made to transfer responsibility for this burgeoning speciality to the Army Service Corps.

Above: Steam Sapper No. 8 with its crew at the School of Military Engineering before leaving for the Ashanti campaign (see page 80). It was not a success for traction due to the narrowness of the roads but served well as a stationary engine back at base.

Above: A Steam Sapper with a train of siege stores at the School of Military Engineering in 1877.

Below: Steam Sappers at work at Frere in South Africa.

Right: Man-lifting kites, capable of operating in wind speeds above the limit for balloons, were taken into service to complement them. The system was developed by an American, Sam Cody, who worked with the RE Balloon Factory for some years. The principle was to hoist a steel cable into the sky using a pilot kite, hold it at the required altitude with lifting kites and then to attach the man-lifting kite with a special harness which allowed the operator to control the basket and to travel up and down at will. Lieutenant Broke-Smith, one of the great pioneers of the technique, reached an altitude of 3,000 feet in this way.

Above: Colonel R. E. B. Crompton, another infantry militia officer whose genius as mechanical and electrical engineer contributed substantially to the Corps' progress in these fields. He had built his own road traction engine at home during school holidays from Harrow. Commissioned into the Rifle Brigade in 1864, he went to India and became ADC to the Commander-in-Chief. Thus he met the Viceroy, Lord Mayo, who showed interest in Crompton's ideas on steam traction. By 1872 he was back in India in charge of four engines and their trains where they proved their value for military transport. Back in England, in 1878 he turned to electrical engineering and founded the company that made his name a household word. In due course Crompton was persuaded to participate in the creation of the London Electrical Engineers. As its commander he went to South Africa with a group of mobile lights which, needing steam power to move them, enabled him also to indulge his first love of steam traction. His expertise was again called upon in the development of the tank (see page 117).

somewhat prematurely dismantled for lack of manpower.

Elsewhere, a balloon section was sent to China for the Boxer Rebellion but, arriving too late, was broken up and its equipment sent to India. There an experimental unit was set up and the equipment was tested on manoeuvres. However, ballooning never really caught on in India due to climatic and logistic difficulties and was abandoned in 1911.

By that time dramatic moves had taken place back in England. A brief flirtation with man-carrying kites was undertaken, but, with the arrival of the internal combustion engine, airships became the logical development from static balloons and many trials were put in hand. But in a period of exceptional financial stringency the British efforts looked pitiful compared with

Above: Major-General Sir John Capper, KCB, KCVO became well-known for his pioneering work in military flying, but had a highly varied career including two years in command of the 24th Division in the First World War in which he led them with success on the Somme (1916) and at Vimy (1917). He joined the Bengal Sappers and Miners on commissioning and, among other active service ventures, built the first road for wheeled transport in the Khyber Pass. In the South African War he raised and commanded a railway pioneer regiment and later three other battalions for guarding railways, and repelled an attack on the Zand River bridge. After his time in divisional command, Capper became involved in the burgeoning tank business (see page 118) and was Colonel Commandant of the Royal Tank Corps until 1934. He died in 1955 aged 93.

the German progress on the Zeppelins. In any case Blériot's cross-Channel flight in 1909 made it clear where the future lay. Fortunately the Duke of Westminster had offered to provide the funds for the Army to buy a Blériot XII and Lieutenant Rex Cammell volunteered to go to France, learn how to fly it and bring it back to England.

Further experimentation with other aircraft, mostly privately owned, then took place before the Air Battalion was formed on 1 April 1911, seven years to the day before the Royal Air Force was founded. By then many more non-sappers had joined the service and within the Corps flying had become a full-time speciality. By May 1912 it was time for the Corps to bow out gracefully and hand the infant arm over to the newly inaugurated Royal Flying Corps, albeit still

equipped with balloons as well as aeroplanes. Many sappers transferred and flew with distinction during the war, three winning VCs: Lanoe Hawker, James McCudden and Mick Mannock.

Above: The airship *Nulli Secundus* was the first such British venture. Its late arrival and 'stick and stick' appearance belied its name in comparison with the Zeppelins, which had flown since 1900. It was built in the Balloon Factory by Lieutenant Colonel James Templer and Mr Sam Cody under the command of Colonel John Capper (q.v.), an enthusiastic innovator in a number of fields. On 5 October 1907 Capper, Cody and Captain W. A de C King flew her on a dramatic demonstration flight from Farnborough round St Paul's Cathedral with a view to raising public interest. The third and last Army airship was ***Gamma***, which flew until 1914 when the Royal Naval Air Service took over all work in this field.

Above: A Blériot XII, the first aeroplane flown by the British Army. Captain Rex Cammell (see below) learned to fly it at the Blériot works in France in June 1909 and subsequently purchased his own Blériot XXI.

Above: Captain Rex Cammell was one of the leading pilots of his day. When asked as early as 1906 why he was applying to join the Balloon Company he replied, 'Flying will be the most important and decisive element in the coming war with Germany and it will begin in that unit.' He was killed in an accident in 1911 while testing a French aircraft of unfamiliar design. He was one of only three officers who had qualified as a pilot of balloons, kites, airships and aeroplanes. Apart from his exploits as a pilot, he had particularly worked on the military applications of flying such as communications, bomb aiming and aerial photography.

Above: The RE balloon pit at Lidsing, Kent (1886), a watercolour by a T. Stammer Day. The pit, on land owned by Major J. L. B. Templer, enabled the balloon to remain safely inflated for long periods.

19 | LAND AND SEA

After decades of complacency post-1815, the British government woke up in the 1840s to a perceived new threat to the security of both the homeland and the Empire, principally from the resurgent French Empire. The alert, sounded jointly by Lord Palmerston and Sir John Burgoyne, resulted eventually in the 1859 Royal Commission into the state of the United Kingdom's fortifications. This was presided over by the sapper Sir Harry Jones (see page 46) and included Royal Navy and Royal Artillery officers in its membership. Not only had no modernisation taken place to accommodate advances in artillery and the emergence of new materials such as iron and concrete; but also steam propulsion now gave potential enemies the capability of landing troops to attack the modestly fortified naval bases from the land.

At home many sappers cut their teeth on the so-called Palmerston forts programme that then transformed the defences of Portsmouth, Plymouth, Pembroke, Portland, Dover, the Thames and Medway approaches and Cork. Much of the implementation of this programme fell into the capable hands of the then Major Drummond Jervois (see illustration). Enormously expensive, the programme attracted much public criticism. By 1884, however, it was nearing completion and Sir Lintorn Simmons was able to refute accusations of over-expenditure (it was inside the budget laid down). He also pointed out that arrangements had been incorporated to allow for the rapidly developing capability of artillery, even while the project was in progress. He further maintained that 'there is not a Power in Europe, which since 1870 has not been occupied in reconstructing the defences of their fortified

Right: Crownhill Fort outside Plymouth. This was the key fortification in the complex laid out to protect the city and dockyard from overland attack. Its design by Captain Edmund Du Cane was approved in 1863; construction began soon after and was completed in 1872, by which time Du Cane had taken over as Surveyor General of Prisons (page 95) in which appointment he was to oversee the design and construction of Wormwood Scrubs.

Above: The Keep of the Dockyard on Ireland Island, Bermuda, a typical example of the dockyard defences of important naval stations throughout the Empire, the work of the Commanding Royal Engineer, Major Andrew Durnford, in 1790. (He was the great-grandfather of Anthony Durnford who was killed at Isandlwana (see page 80). Nine Durnfords over four generations served in the Corps.)

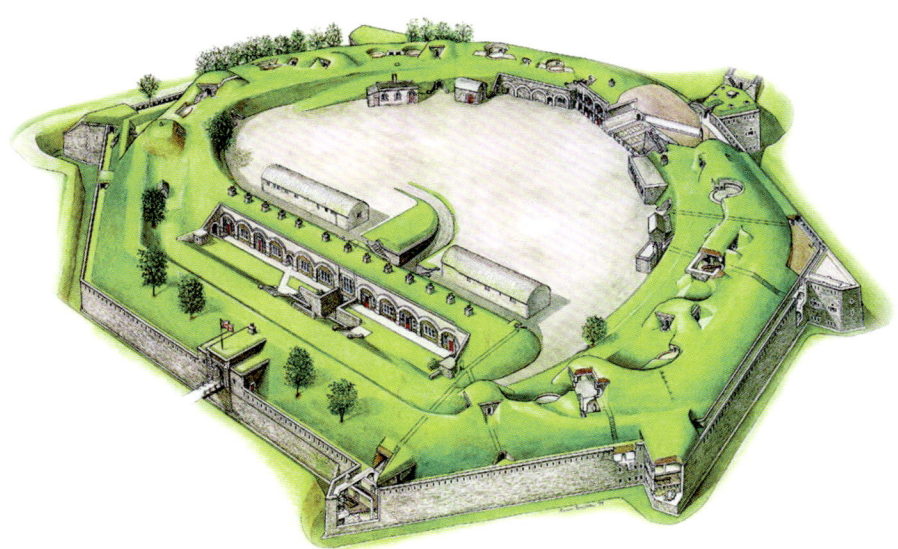

Above: Lieutenant General Sir William Drummond Jervois, CB, GCMG (1821–97), who is best known for his contribution to the development of the fortification of the main naval bases of the country. From 1859 to 1875 he was secretary to the Defence Committee, presided over by the Duke of Cambridge. At this time there was very real concern at the vulnerability of Britain's coast-line to attack from the sea by the French. Earlier in his career Jervois had spent three years on Alderney, seen as a base for a possible attack on Cherbourg, designing and overseeing the construction of the fortifications. As secretary to the Royal Commission on the Defences of the United Kingdom (1859–60) his persuasive argu-ments ensured that the recommendations, though expensive, were accepted in Parliament. All these duties were additional to his responsibilities as Commanding Royal Engineer London District and Assistant Inspec-tor-General of Fortifications. Jervois, whose skills and energies in an earlier appointment in South Africa had attracted attention, now toured the Empire making recommendations on the defence of the colonies. Later he became governor of the Straits Settlements, of South Australia and, after retiring from the Army in 1882, of New Zealand until 1888. On returning to England he continued to write and lecture on fortifications and national defence until his death.

Centre left: Diving in the Medway c.1860.

Bottom left: Nothe Fort Portland under construction, one of the Palmerston forts whose creation was due much to the influence of Jervois.

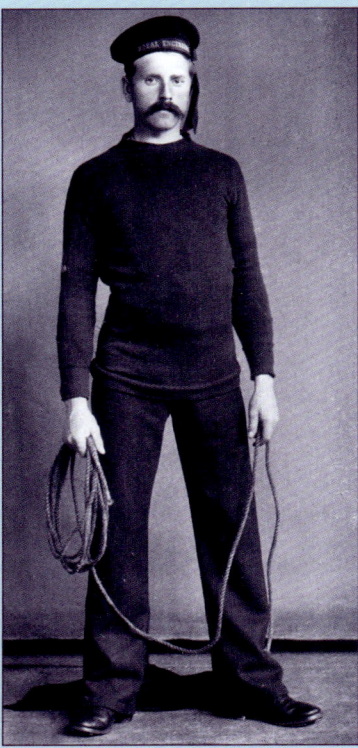

Above: Working dress of the Submarine Mining Service.

Above: Clearing the wreck of the *Royal George*, which had impeded shipping off Spithead since 1782. Colonel Pasley oversaw the project, which began in 1839, with a sapper officer and a detachment of the Royal Sappers and Miners.

Right: snuff boxes made from timber and copper recovered from the wreck and a dessert spoon.

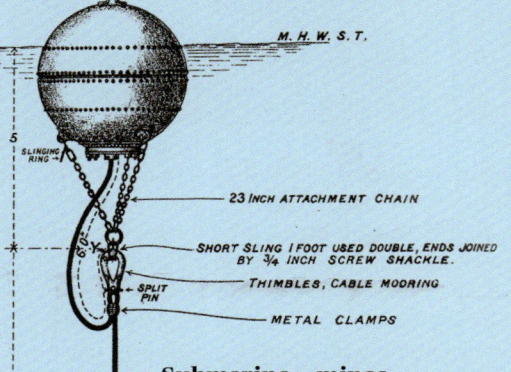

Submarine mines were laid in a carefully charted pattern. They were broadly of two types: contact and observation. In both cases the mines were fired electrically from the shore but in the contact mine a signal was activated by the ship striking the mine thus enabling the operator to identify which mine to fire.

Above: The mine-laying ship *General Skinner*, the largest in the submarine mining service fleet, with four 500lb mines slung alongside.

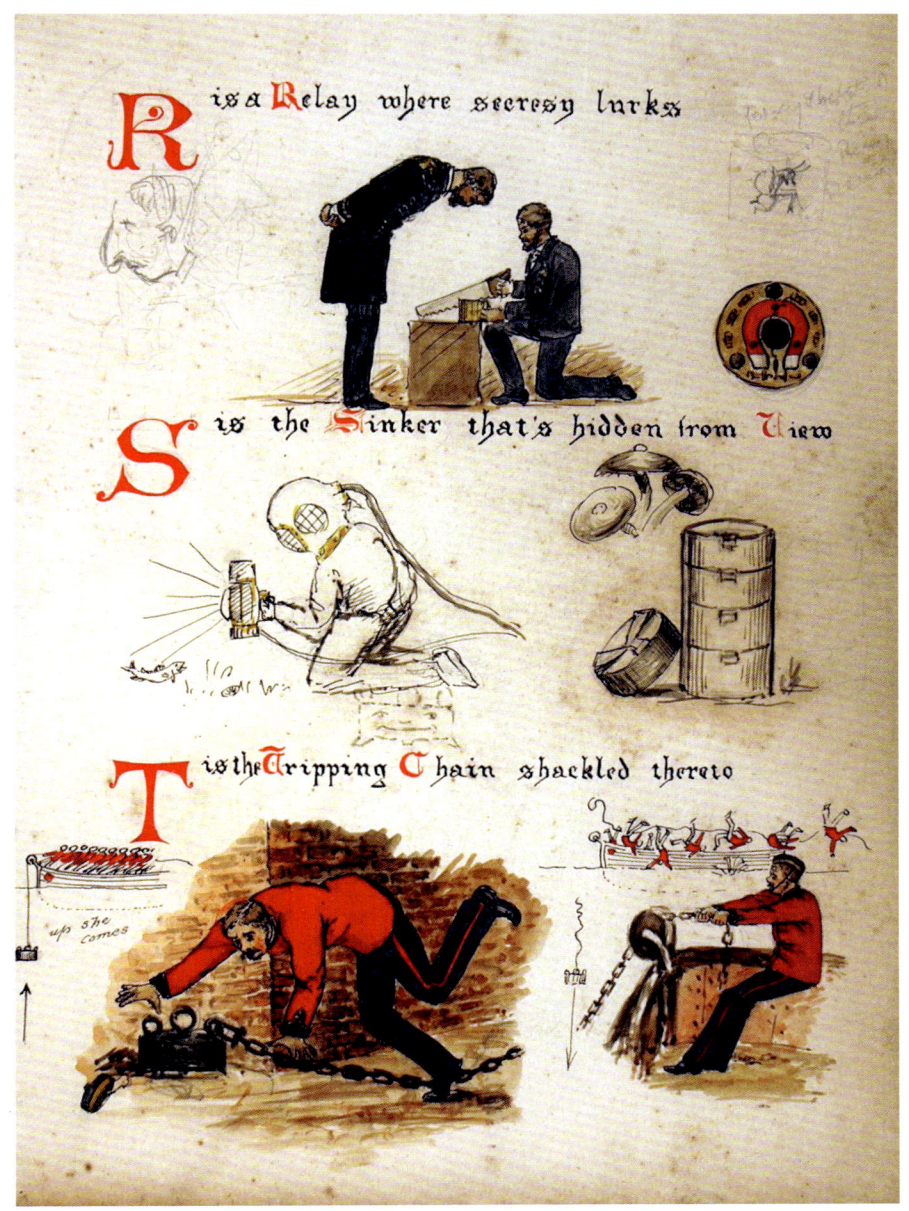

Above: A tunic of a sapper in the Electrical Engineer Volunteers 1900. The unit was formed in 1897 to provide searchlights for harbour protection.

Above: *Submarine Mining, its Stations and Incidents with Illustrations*, a light-hearted look at life in the service by Lieutenant Tyrell-Walker.

places'. It was a policy of deterrence that worked as effectively as its 20th-century successor and possibly exerted an equivalent strain on the national economy.

Indeed there was much concern to upgrade existing defences of British ports world-wide, including the all-important coaling stations. In addition to the threat posed by artillery, a growing concern, was the possibility of attack by torpedo-boat. Another threat was from overland attack by parties landing along unprotected coastlines. Numerous committees sat to discuss matters of policy, design and cost. For example, in 1865, a committee sat under the chairmanship of the Surveyor-General of the Ordnance with such senior officers as the Quartermaster General and Inspector-General of Fortifications as members. Many sapper officers were despatched overseas

Above: A silver lighter in the shape of a 100lb electro-contact mine, presented to Captain A. E. Black of the Lanarkshire Engineer Volunteers, on the occasion of his marriage.

to report on existing defences, and gradually improvements took place.

This work coincided with the realisation that underwater charges* could be dramatically effective against warships. Fortunately the Corps was well advanced in these matters since the 1830s when Colonel Charles Pasley, then commanding the Royal Engineer Establishment, evolved the techniques of firing underwater charges, and personally became the first service diver in history when he experimented with the Siebe and Gorman apparatus. Expertise was further developed with the diving-bell and numerous tasks were undertaken, often in conjunction with the Royal Navy, of which the clearance of the wreck of the *Royal George* (q.v.) was but one example. Thus began the Corps' long tradition of diving.

Although Pasley had invented an electric detonator, it was not until the 1870s that controlled

firing of underwater mines by electricity became a practical possibility. Soon submarine mining, in conjunction with artillery, formed an integral part of all major fortifications. The Brennan torpedo (q.v.) was added in eight key sites. Defence searchlights were introduced, giving 24-hour cover of the minefields. These were also used to illuminate battlefields on land, including in India. Sapper expertise in this speciality led eventually to the Corps commitment to anti-aircraft searchlights although these were transferred to the Royal Artillery in the Second World War (see page 147).

The new submarine mining service that then grew up became a substantial part of the Corps. By the time of its demise the service numbered 5,890 men of whom 2,000 were regular, the rest militia, volunteers and reservists. The Corps remained responsible for the service until 1905 when the Royal Navy took over responsibility for all sea mines and torpedoes. Diving, however, remained an essential part of the Corps' life.

*These were generally known as 'torpedoes' until later in the century when static controlled charges were dubbed 'submarine mines', borrowing a term from land warfare, and 'torpedo' was reserved for powered explosive devices.

Above: A Brennan Torpedo, the only one surviving. This was the world's first dirigible torpedo. It was developed for harbour defence, complementing submarine minefields. It was launched from a shore station. Internal drums carried wires that were withdrawn by a powerful engine ashore. This rotated the drums whose rotational energy was converted to forward power by means of propellers. The mechanism was highly complex and so secret that some details are still not clear.

The Brennan had a range of about one mile, carried a warhead of 220lb of wet guncotton and travelled through the water at about 27 knots. It was never used on operations, but its effectiveness was proved by regular practice, undertaken throughout the 1890s by the Brennan teams.

Louis Brennan, its inventor (right, a caricature by Herbert G. C. Allen), was an Australian of Irish origins who brought his ideas to England in 1879. After development work and trials, a lucrative agreement was arrived at for the production and exclusive use of the torpedo by the Royal Engineers. Brennan was a prolific inventor whose other brain-children included a gyro-stabilised monorail train and an autogyro type of flying machine.

20 | WIRE AND WIRELESS

Signals as a Royal Engineer speciality began with the Crimean War telegraph system: 21 miles of line with eight stations connecting the principal headquarters and connected (by contract) to the submarine cable that had been laid across the Black Sea. No specialist units were established, however, until government anxiety at the Franco-Prussian War (1870–1) prompted the authorisation of a telegraph troop – C Troop, with 24 wagons, twelve of which each carried three miles of wire.

At the same time the use of sappers to help work the government telegraph system under the Post Office was approved and 22 and 34 Field Companies were allotted to this

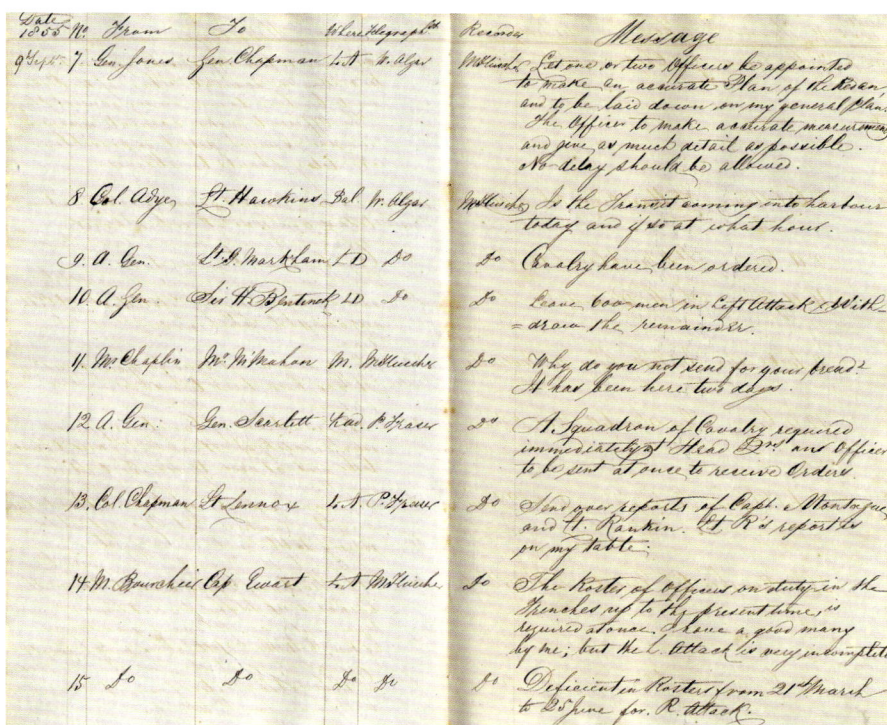

service. Some individual soldiers so employed had already had experience in the Abyssinian campaign (1867–8) (see page 75). Similar steps were taken in India, lent urgency by the 1857 Uprising, and throughout the 1860s and 1870s, sappers took a leading role in setting up and running the overland line to Europe through Persia.

Above: An extract from the telegraph logs used in the Crimea. The entry is for 9 September 1855, the day after the Russians evacuated Sevastopol after the final allied attack.
Examples: Serial 7, from General Jones to Colonel Chapman: 'Let one or two Officers be appointed to make an accurate plan of the Redan …'
Serial 12, from the Adjutant General to General Scarlett (commanding the Heavy Brigade: 'A Squadron of Cavalry required immediately at Headquarters …'

Above: RE signallers fixing telephone lines near Fricourt, 1916. The telephone was introduced in 1876, but was slow to catch on apart from in the Submarine Mining Service. The Boer War (1899–1902) proved its indispensability. (IWM Q4137)

Above: A Mark II Long Wave Tuner used in early First World War wireless. It was invented by Major H. P. Lefroy in 1914.

Above: A signal lamp from Napier's 1867 Abyssinia campaign. Visual signals were the main means of communication in the field throughout the 19th century although military provision of telegraph for rear links started in the Ashanti War.

Above: The crest of the Telegraph Battalion RE, formed in 1885, was the figure of Mercury, popularly 'Jimmy', which was adopted by the Royal Corps of Signals on their formation in 1920. The crest shown here on a bass drum was the design of Major Charles Beresford, the first commanding officer.

Right: A mobile wireless station supplied to the Army under a Marconi contract for trials in 1901. The antenna could be lowered to the horizontal position for travel.

By 1885 experience in war had borne out the necessity of telegraph and a battalion of two 'divisions' was formed, serving overseas with distinction (see page 83). At this time visual signalling by flags and lamps had developed as an Army-wide service with a training school at Aldershot. The post-Boer War Haldane reforms led to a combined Telegraph and Signal Service under the Corps in 1906. At the same time divisions of the field army were allotted a telegraph company each to add to their two field companies of sappers. By 1911 the technical possibilities and military potential had burgeoned. 'Signal' had replaced 'Telegraph' in unit titles the previous year. A War Office committee considered forming a separate signals corps, but decided to leave the service as a speciality within the Royal Engineers. However, the Telegraph School at Chatham closed in 1913 and the Army School of Signalling in Aldershot took over all training in both visual and electronic skills.

Telephones had first made their appearance with the submarine mining units. In South Africa they were deployed by the electrical engineers, but thereafter were subsumed into the telegraph units. Likewise, wireless telegraphy was experimented with during the Boer

Above: Carrier pigeons were used extensively in the First World War. By 1918 there were 20,000 birds in service requiring some 380 pigeoneers. At least one of these was awarded the MM for his services. As in this scene, pigeons had to take the place of telephones when the vulnerable lines were broken. (IWM Q58208)

War* and afterwards by the early pioneers of flying.

During that war signals, mostly by telephone, developed into the essential nervous system of the Army, as it is today. In 1914 there were 31 regular and territorial units in the Corps. By the end of 1918 that number had risen to 589 in 77 different types of unit ranging from line construction and carrier pigeon to messenger dogs and wireless telegraphy. The war led to some urgency in experimentation with wireless communication particularly for ground-to-air links, which it was thought might help guide aircraft to tackle night-flying Zeppelins. The first man-transportable set was also produced, with a range of fifteen to twenty miles. However, throughout the war the main methods of communication remained the telephone and telegraph. Much work was also done on interception and location of enemy stations by direction-finding and conversely on security. Major A. C. Fuller devised a telephone protected against intercept by induction methods. The 'Fullerphone' entered service and remained a standard equipment for many years.

On 27 June 1920 the Signal Service of the Royal Engineers became the new Corps of Signals, the Royal title being granted the following August.

*As early as 1809 Colonel H. H. Austin had accompanied a Royal Navy expedition operating off the Persian Gulf using wireless to intercept gun-running from Muscat to the North-West Frontier.

21 | MAKING TRACKS

Above: A 'tankette', prototype of the Bren Gun Carrier, an early demonstration of a tracked, armoured vehicle. Martel (see page 118), whose head can be seen in the photograph, built this machine himself.

Below: A Mark I 'male/female' tank, the first type ever to go into action, on 15 September 1916 during the battle of the Somme. The artist was Major Walter Keesey, who fought in both France and Italy, winning the MC.

The idea of an armoured cross-country vehicle capable of penetrating wire obstacles and surmounting trenches was conceived in February 1915 by a group of men, eventually forming the 'Landship Committee', working under the dynamic authority of the First Lord of the Admiralty, Winston Churchill. Colonel R. E. B. Crompton (see page 108) also appeared at this stage of the tank story by providing and developing under contract some early tracked vehicles. A difficult gestation was supervised by the sapper, Colonel Ernest Swinton (q.v.), Secretary to the War Committee, whose creative mind and perception of technical possibilities, saw the infant through its birth and its formative years, christening it a 'tank' when a deceptive name was sought for reasons of secrecy. At the beginning of the war Swinton had been despatched to France by Kitchener to act as the official war correspondent with the BEF, based at GHQ. It was then that he grasped the potential of such a device and it was fortunate that by July 1915 he was in a position to influence decisions.

Tanks first went into action in September 1916 in the Somme battle, losing the element of surprise that Swinton had advocated but gaining early experience. The Flanders mud showed up the limitations to their mobility during Third Ypres but in November 1917 the more solid chalk of Cambrai proved their potential. By that time they had been formed into the Tank Corps under the sapper, Major General Hugh Elles (see page 118), with Colonel J. F. C. Fuller (a former infantryman) as his principal staff officer and Major Giffard Martel (q.v.) as GSO2. Later 'Boney' Fuller became one of the foremost exponents of armoured warfare doctrine and Martel the architect of the Royal Armoured Corps during the Second World War. Another sapper, Major

Above: Major General Sir Ernest Swinton, KBE, CB, DSO, whose insight and determination had prominent influence in the creation of the tank. His early career took him to India and South Africa and he became well known as an acute observer and writer on military affairs. In 1914 Kitchener plucked him out of his job as Deputy Director of Railways to be the official correspondent for the BEF, known as 'Eyewitness'. In July he became Acting Secretary to the Committee of Imperial Defence and a member of the Tank Supply Committee. He commanded the first tank unit and oversaw its induction to war on the Somme, but his age ruled him out of further command and he rejoined the War Cabinet Secretariat for the rest of the war. Retiring from the Army after the war, he pursued a second career as a civil servant mostly in aviation. (Portrait by Jane Corsellis)

General John Capper (see page 108), had entered the scene in 1917 as Commandant of the training depot of the tanks, then called the Machine Gun Corps Training Centre. Capper became Director General of the Tank Corps at the War Office and later the first Colonel Commandant of the Royal Tank Corps.

For the Corps, the arrival of the tank led to an immediate reappraisal of bridging equipment, stronger modifications of existing designs coming into service. From the Tank Corps' point of view, the need for specialist engineer tanks had become obvious from the first operations. After some improvisations in the early days it was decided to form three special tank bridging battalions, each with 48 tanks and appropriate bridges and pontoons. Peace caught up with this proposal and only one battalion was formed, the first mechanised Royal Engineer unit, and that was soon reduced to become the Experimental Bridging Company, Royal Engineers at Christchurch. Thus was founded the home of British military bridging.

Left: General Sir Hugh Elles, KCB, KCMG, KCVO, DSO (1880–1945), who, as a lieutenant colonel serving on the staff on the Western Front, was selected to command the 'Heavy Section of the Machine-Gun Corps', which became the Tank Corps (see page 125). He led them into the battle of Cambrai, the first use of mass armour in history. From his appointment as Director of Military Training in 1926 to that as MGO in 1934, he devoted much effort to promoting a belief in mechanical warfare against a background of political apathy.

Left: Lieutenant General Sir Giffard Martel, KCB, KBE, DSO, MC, whose experience as a GSO2 under Sir Hugh Elles led to a post-war involvement in tanks and their operational use, including in the engineering role. He was commissioned in 1909 and went to France with 9 Field Company before joining Elles. After the war he commanded the Experimental Bridging Company at Christchurch, effectively founding the future MEXE. Later he was further involved with mechanisation and while in a staff appointment designed and built a light tank of his own in a borrowed workshop. In the Second World War he commanded the 50th Division and took them to France in 1940. On return to England he was appointed Commander Royal Armoured Corps. He retired in 1944.

Left: The Mark IV tank, the first really effective armoured fighting vehicle, whose success at the battle of Cambrai proved the battle-winning potential of tanks. Camouflage by 'dazzle painting' the tanks was undertaken by Lieutenant Colonel J. Solomon, one of the artists specially commissioned into the Corps.

Below: The 21-foot Canal Lock Bridge carried by the heavy Mark V tank, an experimental method of crossing canal locks, typifies the work of the then Major Giffard Martel at Christchurch in the years immediately following the 1918 Armistice. (IWM T18)

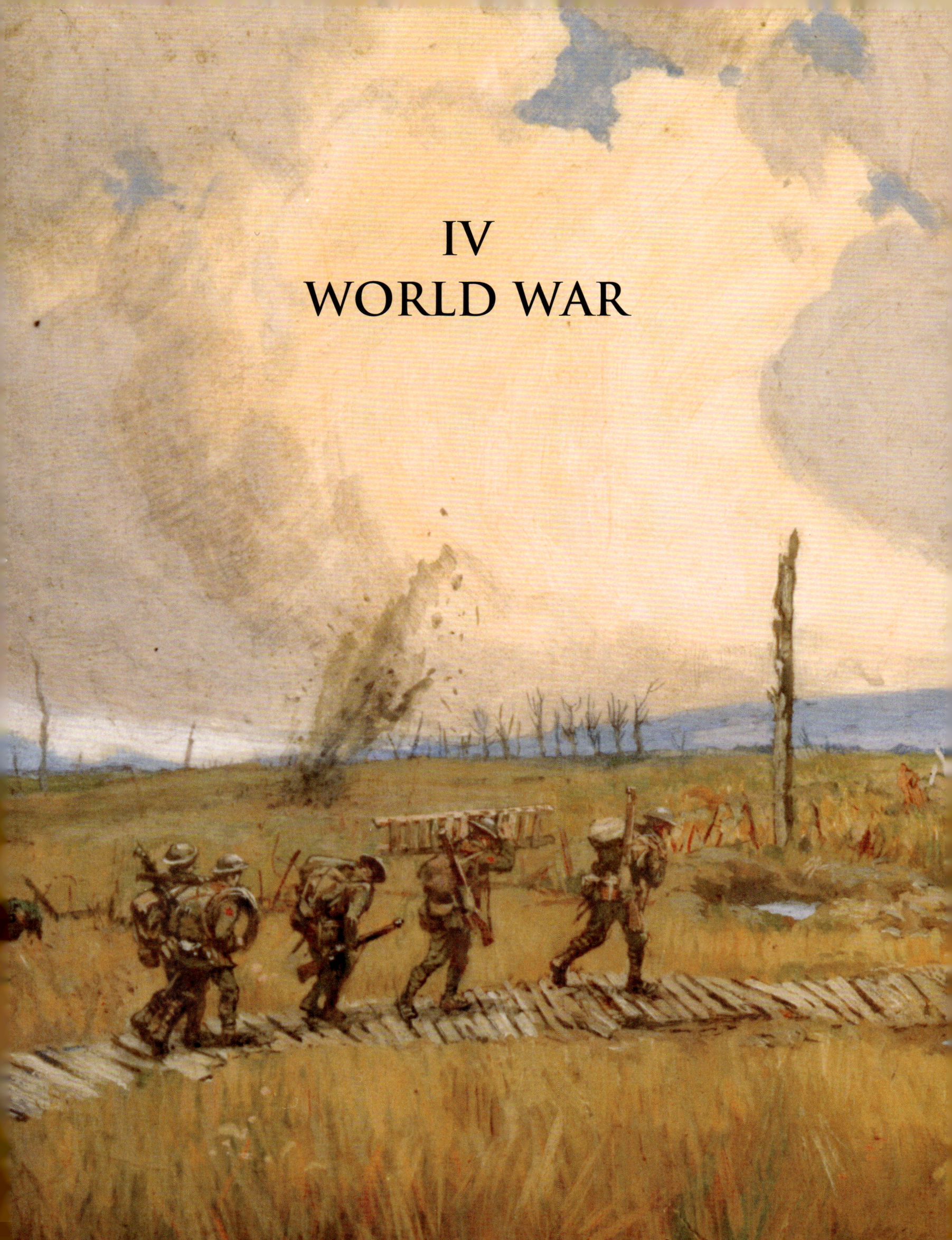

IV
WORLD WAR

THE WESTERN FRONT 1914 AND 1915

The first sappers to confront the enemy on 23 August 1914 were the two regular field companies of each of the forward divisions, 56 and 57 in the 3rd Division and 17 and 59 in the 5th, forming II Corps of the British Expeditionary Force holding the line of the Mons–Condé Canal. Five German corps were massed on this front in implementation of the Schlieffen Plan. By that plan the German high command hoped to outflank the Allied forces and seize Paris, force a French surrender and then advance east against Russia. Although overwhelmed, the British Expeditionary Force fought magnificently and, in its retreat from Mons, made a major contribution to wrecking the German grand design, which finally came to grief at the battle of the Marne.

Their achievement owed much to the reforms that had followed the South African War during Lord Haldane's tenure as Secretary of State for War (1906–12). The Corps emerged from these with the divisional engineers of the six expeditionary force infantry divisions (two field and one signal company each) and the field squadron and signal troops of the Cavalry Division. In addition there were the field sappers of the fourteen new Territorial Army divisions plus twenty-three works and nineteen electric light companies. From this modest base the Corps increased tenfold during the years of the war, as part of the colossal expansion of the whole Army under Kitchener who took over as Secretary of State on 6 August 1914. At once he declared that the country must prepare for a war to last three years for which seventy divisions must be raised. Scarcely anyone believed him, but the authority that he carried won everybody's cooperation.

The British Expeditionary Force plugged the gap left by the overstretched French armies and the battles flowed south to the Marne. The French Commander-in Chief, General Joffre

Above: A sense of the desolation of **the Menin Road,** the main route out of Ypres into the Salient, and its surrounds is caught here by the camouflage artist Captain W. F. C. Holden.

Left: The Bangalore Torpedo, the invention of Major R. L. McClintock of the Madras Sappers and Miners, typical of the many improvisations resorted to in the early days of the war. Lengths of metal tubing, filled with explosive, were pushed through barbed-wire entanglements and exploded to cut the wire. In this case the original improvisation was eventually adopted officially.

Below and on subsequent pages: A field company on the move, a frieze in silhouette by Sapper W. Carr.

Above: The Military Cross and Military Medal were instituted in 1914 and 1916 respectively to provide gallantry awards for junior officers and warrant officers (MC), and non-commissioned officers and soldiers (MM). Previously the only awards for acts of gallantry not qualifying for the Victoria Cross were the Distinguished Service Order and Distinguished Conduct Medal, which were generally available for achievements not necessarily in presence of the enemy.

(a former Engineer) manoeuvred to throw the Germans into confusion, back north to the Aisne. The Germans hung on and, as both armies tried to outflank each other, by mid-October both sides were digging in along what was to become their line of confrontation for most of the rest of the war. Around Ypres desperate German efforts to break through were denied by the remnant of the British Expeditionary Force, Kaiser Wilhelm's 'contemptible little army', at the battles of First Ypres (October and November 1914) and Second Ypres (April and May 1915). It was a remarkable feat in which the sappers had played their part, winning five VCs. Fighting as infantry was commonplace. Two of the VCs had been in that role, three of them in typical sapper front-line situations. On 11 November 5 Field Company became particularly embroiled in the fighting at Nonne Bosschen, near Ypres, for which they were awarded seven DCMs.

By this time the Empire had rallied to the cause. India had sent two divisions, which arrived in Marseilles in September. They were filling a gap in the British line by mid-November. Their intervention at this critical juncture not only prevented a German breakthrough but allowed time for the first of the Territorial divisions to be prepared to join the line as complete formations rather than as individual reinforcements. The Canadian Expeditionary Force and Australian Imperial Force both left their respective countries in October 1914 and the Anzac divisions started forming up the following January.

These early days were hectic ones for the sappers. Trench work, wiring, providing water, building bunkers and redoubts, bridges to move the guns forward and all the associated logistics and innumerable small tasks to help the infantry fight were on their agenda. Improvisation with local materials was fundamental to

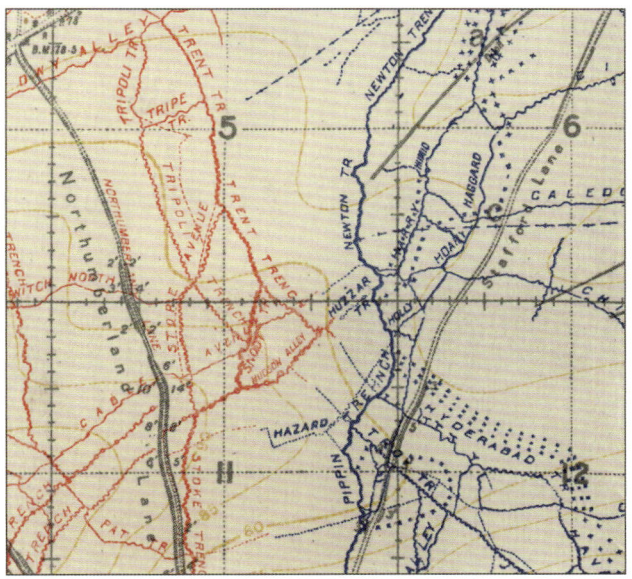

Above: Prodigious numbers of **trench maps** at numerous different scales were produced by RE Survey. At one time some eight tons of them were shipped to France daily. Illustrated is a section of the 1:20,000 sheet 51B NW dated 8 July 1918 showing British (red) and German (blue) trenches in the area of Bailleul to the north of Arras.

Above: Lieutenant General Sir Aylmer Hunter-Weston, KCB, DSO, shown wearing the black mantle of a Knight of Justice of the order of Saint John of Jerusalem. Hunter-Weston made his mark as a junior officer on the North-West Frontier and in the South African War (see page 88), when he was serving with the newly formed Field Troop part RE in which all the sappers were mounted. This led to the recognition of the mounted Field Squadron as an integral part of the Cavalry Division. In the 1914–18 war, he commanded 29th Division and then VIII Corps at Gallipoli. He remained VIII Corps commander on the Western Front including at the crucial Beaumont Hamel sector on the opening day of the Somme battle on 1 July 1916. Throughout his career he displayed singular powers of leadership. He had a kind but forceful personality, and his somewhat eccentric manner only served to endear him further to the troops as 'General Hunter-Bunter'. After the war he sat in the House of Commons as Conservative MP for North Ayrshire.

AFFIRMATION

'Let not the world mistake us. Should any outside danger threaten us we will stand shoulder to shoulder round our mighty mother, England, and her enemies will find us arrayed in solid phalanx by her side, ready to meet any danger and render any sacrifices for the sake of the great and glorious Empire of which we are proud to call ourselves citizens.'

Speech by Sir Gangadhar Chitnavis in the Viceroy's legislative Council, August 1914. (Lieutenant Colonel E W. C. Sandes, *The Military Engineer in India*, vol. I, p. 475)

Right: German OP La Bassée by the camouflage artist, Captain A. R. Harker, who must have enjoyed recording the efforts of one of his German opposite numbers.

Right: Trench tramways were introduced as early as 1915 to move stores forward and spoil from tunnelling and trench work to the rear. More elaborate systems began to appear in 1916. (See also page 134.) (IWM Q7908)

success. A workshop manufactured all manner of stores, makeshift bombs, grenades, periscopes and trench mortars. Two-foot gauge wooden tramways were built, foreshadowing the light railways of the future. Such was the pressure that during the winter a third (territorial) field company was added to all divisional engineers.

On the wider scene, Survey in its most pro-active role became indispensable to success. In November 1914 the 1st Ranging Section RE joined the 8th Division Artillery to develop target location methods with the help of the Royal Flying Corps and the French national triangulation system. In due course these units became responsible for the re-mapping of the entire British Front. Sound-ranging and flash-spotting developed naturally from these activities. There were five survey battalions on the Western Front by 1918 and 5,000 men committed to this service, a relatively modest manpower bill for results of such consequence.

The new year (1915) brought disillusion as the Allies went on the offensive in Artois (the British at Neuve Chapelle, Festubert, Aubers Ridge and Loos, in support of French endeavours elsewhere), coming tantalisingly close to breaking through, but failing for lack of artillery and the means to exploit success. As the spectre of stalemate loomed, new ideas were conceived that were to engage the sappers in the years ahead: mines, tunnelling, gas and tanks all made an appearance either experimentally or on the drawing-board at this time. But the principal pressure for a new initiative was on the wider strategic scale, in the shape of the Gallipoli campaign.

GALLIPOLI AND SALONIKA

Winston Churchill, as First Lord of the Admiralty, had advocated that by striking at Turkey through the Dardanelles and capturing Constantinople the Central Powers' whole strategy would be undermined and pressure taken off both the Western Front and Russia. At first Kitchener was unwilling to spare land forces for this enterprise and Churchill believed it could succeed through naval action alone. It nearly worked; but in early March as the fleet became more and more bogged down in minefields, Kitchener acceded to a force being sent to land on the Gallipoli peninsula on 25 April 1915. Under the command of General Sir Ian Hamilton, it comprised 75,000 men. More than 30,000 were Australian and New Zealanders (Anzacs)* and there was the Royal Naval Division of 10,000, a French division of 17,000 and the slightly stronger British 29th

*Anzac Cove, the landing point of the Anzac force, was on the west coast, some fifteen miles north of Cape Helles where 29th Division landed.

Division under the command of Major General Aylmer Hunter-Weston (see pages 88 and 121).

There were both disasters and triumphs in these landings, but the Turks reacted with alacrity and everywhere the action resolved itself into a war of attrition. In July five new divisions arrived, bringing the total in the theatre to thirteen, more than half the number in France. Coordinated attacks to support a new landing at Suvla Bay were made. Supreme gallantry with awful losses took the survivors of the leading battalions to the top of the central ridge of the peninsula from which they could tantalisingly see the Dardanelles, but this superhuman effort fizzled out. Once again a military impasse was reached. After much military and political recrimination and changes in command, Kitchener himself went out and agreed that withdrawal was the only realistic option. For once fortune smiled and in superb weather a meticulously planned withdrawal operation cleared the whole peninsula, less the southernmost tip, before Christmas. The last troops left on 8 January 1916.

The Corps *History* (Vol 5, p 123) records that 'Trench warfare at Anzac was immensely interesting to the sapper.' So it was throughout the operation, with the problems presented by the exceptional terrain. Improvisation had to make up for lack of materials. Priorities were water supply, field defences and the piers, beach crossings and roads and tracks on which depended both logistic support during operations and the safety of the force in the withdrawal operation. Extensive mining developed into an underground war pre-dating much that was to occur on the Western Front. The bases on the offshore islands of Lemnos and Imros also absorbed much engineer effort.

Even before the decision had been taken to evacuate Gallipoli, the German and Austrian armies swept into Serbia and threatened Greece. A force had to be found to go to Salonika and was scratched together based on one each of the French and British Gallipoli divisions and other elements designed to bring the port into use. This 'sideshow' war built up into a tiresome commitment in an unpleasant climate, engaging by the end of the war six French and six British divisions, another six from the Serbian Army plus contributions

from Russia and Italy. It cost the British 24,500 battle casualties and 162,500 were admitted to hospital with malaria.

As if Gallipoli and the Balkans were not enough, farther east the Turks were mounting an even greater threat.

MESOPOTAMIA

Turkey's entry into the war threatened British oilfields in Persia. A force of largely Indian Army troops was built up, landed in Basra and initially made brilliant progress forcing the enemy back up the Euphrates (under the sapper Major General George Gorringe) and Tigris. The Tigris force under Major General Townshend made a famous dash towards Baghdad but was driven back to Kut-al-Amara where it was besieged from December 1915 to April 1916. Efforts to relieve Kut, by the Tigris Corps under the sapper Lieutenant General Sir Fenton Aylmer, VC (see page 57), was defeated by stout enemy resistance, inadequate force, hideous logistic problems and shortage of time. More than ten thousand

Above: Gully Ravine, in the Cape Helles area of Gallipoli. Here 42nd Division engineers built dug-in winter quarters for the reserve brigade during the autumn of 1915.

Above: Wiring was a frequent task for the field companies, often conducted under more hazardous conditions than in this good-humoured interpretation by Sapper W. Carr.

LOCATION OF RE GRAVE, RAILWAY WOOD

Above: A section of a battlefield panorama showing a sector of the Ypres salient looking north from Rifle Farm close to 'Hellfire Corner' on the Menin Road. The buildings of Hooge Château are on the far right. Thousands of these panoramas were produced by military survey to provide up-to-date intelligence for commanders. They were created manually or photographically (as illustrated), mostly the latter, either to provide a general view of the ground or specific target information. The camera work was hazardous, the operator having to find a position in a forward area with a good view, the camera to be mounted on a special levelled stand with graduated arcs and pointers to assure that each separate photograph was correctly aligned with its neighbour. All sapper officers were taught field sketching and many became skilled artists.

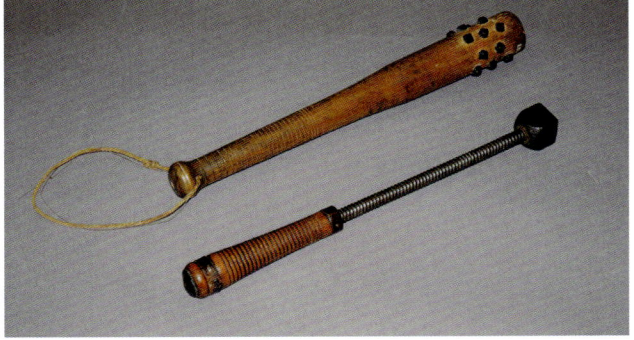

Left: Trench weapons. These maces are typical of the type of weaponry produced in the RE Workshops in France in the urgency of the early days of the war. (See also page 132)

Left: Indian sappers recovering Turkish war material from the River Tigris. (IWM Q24585)

men were marched into captivity where 4,000 of them died. Some 40,000 casualties had been incurred in the attempts to relieve them.

THE WESTERN FRONT 1916

While Britain and France were planning the 'Great Push' in Picardy, where the River Somme divided their two armies, the Germans were poised to pre-empt such a move by attacking Verdun, not so much to break through as to suck the French Army into a killing ground and destroy it. Their blow fell in February. It all but achieved its aim in the longest battle of the war (21 February to 16 December). The burden of the Somme battle therefore fell on the British with their newly raised 'Kitchener Armies'. These two battles – Verdun a holocaust in which the French eventually prevailed but at irreparable loss, and the Somme in which 60,000 British soldiers were lost (19,000 killed) on the first day, 1 July – proved the strength of well dug-in trenches protected by wire, even against massive artillery concentrations.

By the end of the year little ground had been gained on either front. Allied resolve and sacrifice in these terrible struggles had, however,

dealt the enemy a mortal wound from which there would be no recovery. The Germans withdrew behind the massive pre-prepared defences of the Hindenberg Line early in 1917, intending to concentrate their efforts on bringing the war to an end on the Russian Front.

By August 1916 the Corps had built up from its pre-war strength of seventy-eight regular and 100 TA units to more than 900 of which 550 were on the Western Front. About 450 of these were field and signals units.

1917

Allied hopes for success in 1917 hung on the big coordinated spring offensives, by the British and

Below: Sanctuary Wood, in the Ypres Salient, a watercolour by Lieutenant Richard Cooper, another of the camouflage artists. Sanctuary Wood was in the Ypres Salient south of the Menin Road and was very much in the eye of the storm throughout the war. For example Captain John Glubb's diary entry for 2 March 1916 reads: 'In Sanctuary Wood we created a diversion by bombs, trench mortars, blowing mines etc and got badly shelled in return.'* The picture is signed 1919.

Canadians at Arras and the French in Champagne. At Arras, a fine start, thanks largely to the enlightened use of tunnels in the approach, eventually foundered on the outer reefs of the Hindenberg Line. However, the Canadians had captured Vimy Ridge and the main aim of drawing off the Germans from the French front had been achieved. This became strategically crucial when the French offensive failed so disastrously that mutiny raised the spectre of a German breakthrough.

By this time the wider strategic picture had altered dramatically with the collapse of the Russians on the Eastern Front in March and the American declaration of war on 6 April.

The British focus now moved to Flanders, the front that Haig had always believed offered best prospects of success. In June the battle of Messines (see page 128) raised hopes of a break-

EXPANSION SOME STATISTICS OF THE EXPANSION OF THE CORPS from 1914 to 1918		
Types of units	24	59
Officers and soldiers regular and TA (excl. Transportation)	25,000	230,000
Transportation	500	62,500[†]
British divisions	6	70
Units (incl TA)	205	1,832

[†]Plus a further 50,000 colonial and dominion.

through but these were to sink beneath the mud that brought Third Ypres (Passchendaele) to a halt in November. Casualties were almost on the scale of the Somme.

However, the Allies had one more card to play in November when nearly 400 tanks were launched by Third Army in the first mass armoured attack, at Cambrai (see also page 118). Dramatic success was followed by disappointment as the German counter-attacks forced back the British Third Army almost to their original start line, for want of reserves to exploit their advantage.

The year ended with glimmers of more permanent success against the Turks in Mesopotamia, where Baghdad had been occupied on 11 March 1917, and in Palestine, where Jerusalem fell on 9 December. Against this, the Italians were pushed almost to the point of capitulation by a catastrophic defeat at the battle of Caporetto (24 October to 12 November). The Allies responded by sending eleven divisions to bolster their allies, five of which were able to be withdrawn in the new year, in time to meet the onslaught that fell upon the British front in March 1918 (see page 135).

*A complete set of the original diaries of the future 'Glubb Pasha' (see also page 141), written in army issue notebooks, is held by the RE Library.

23 | UNDERGROUND WAR

With the pattern of trench warfare already set before the end of 1914, the idea of tunnelling under enemy lines was obvious to both sides. But the British field companies had neither the skill, equipment nor manpower to make serious progress in the art. Urgency was injected by General Sir Henry Rawlinson, commanding IV Corps, with his plea for a specialised battalion to be raised. This followed the first successful German operation near Givenchy, on 20 December 1914, when the Sirhind Brigade was shattered after ten small mines had been fired under their forward trench line.

Earlier that month, the remarkable Major J. Norton-Griffiths, MP, had foreseen the significant effect that tunnelling might have on the Western Front. He managed to get his ideas before Kitchener, who was at that time at a loss as to how best to meet the anxious calls from Flanders, and offered to divert men to the cause from a project he had won for a tunnelled drainage system in Manchester.

Eight tunnelling companies were immediately authorised, but so great was the demand that a further eighteen British and seven Commonwealth, three each from Canada and Australia and one from New Zealand, were in theatre by 1916. According to the Corps History the numbers in this vast force were not '… far short of the total of other engineer effectives engaged in the front line in France'. The first of these men were working in the front line, in uniform and fully kitted, within two weeks of enlisting.

The underground war was no side-show, but matched the eternal battles for supremacy above ground, conducted in hundreds of battalion areas. For example, on a busy day in June 1916, 227 mines were fired along the whole British front, 101 by the British and 126 by the Germans; 79 and 73 respectively were fired on the First Army front. The tunnellers' most urgent business was to counter the German offensive mining operations. The first successful countermine was

Above: Major J. Norton-Griffiths, MP, the inspiration behind the British response to the German initiation of the underground war in 1914 (see text). John Norton-Griffiths was a colourful character of exceptional determination and ability to achieve results. Before the war, he was well-known as a wealthy entrepreneur, a civil engineer with his own internationally renowned company. His military experience included service in South Africa with the 2nd Division and later as Captain and Adjutant of Lord Roberts's body-guard. He was also instrumental in raising a Yeomanry cavalry unit before the war broke out, paying much of the cost of billeting and equipment from his own pocket.

Right: German Mine Listening set.

Above: The Tunnellers' Friends, a replica of the tunnellers' memorial in St Giles Cathedral, Edinburgh, depicting the mice and canaries taken into the tunnels to give early warning of accumulations of lethal gases.

Left: Martha House, a British dug-out begun in December 1917, photographed in 1998.

Left: Sapper Willliam Hackett, VC was a miner with twenty-three years' experience in the Nottinghamshire coalfields when he enlisted in October 1915. He joined 254 Field Company in France less than a month later. On 22 June 1916 he was in a party working in Shaftesbury Shaft at Givenchy when a fall resulting from a German counter-mine cut them off. A rescue team reached them through a small opening and three men struggled out to safety. Sapper Hackett and Private Collins of the Welch Regiment, who was badly injured and could not be brought through the small aperture, remained. Hackett was urged to follow before a further fall of the loose earth could prevent any chance of escape. He declined with the words, 'I am a Tunneller. I must look after my mate.'

Below: His VC.

Above: The Military Cross and war medals of Lieutenant W. R. Cloutman, who was killed in August 1915 in a fall while saving the life of a sergeant in 178 Tunnelling Company. Cloutman had led the first successful countermine operation, in the Ypres area, by a party of 171 Tunnelling Company early in 1915. His brother, Major B. M. Cloutman, won the last sapper VC of the war, on the Sambre-Oise Canal in November 1918.

Left: A British officer using a geophone.

MESSINES, 7 JUNE 1917

'In a flash the monsters of destruction lurking below the ground were goaded into wakefulness. With a stupendous paroxysm they shook themselves free, convulsing the earth for miles around ... It seemed as if the Messines ridge got up and shook itself. All along its flank belched rows of mushroom-shaped masses of debris, flung high into the air ... This was a man-made earthquake, a spectacle out-majestied only by Nature in her most ungovernable moods.'
Royal Engineer tunnellers had been preparing for this event for over two years. On the efficiency of their work hung the success or failure of the attack and hence the lives of thousands of men. This was the greatest offensive mining operation of all time. Many of the nineteen mines, each containing 20,000 to 100,000 lb of ammonal, laid 50 to 100 feet deep and at the end of galleries some 1,500 feet in length, had been emplaced more than a year earlier, their condition being painstakingly maintained against the ravages of the climate, enemy action and accidental falls of earth.

Above: Mine explosion at the Hawthorn Redoubt ten minutes before the main attack on the Somme on 1 July 1916. (IWM Q754)

Far right: A tunnelling plan drawn up by 177 Tunnelling Company at Railway Wood in the Salient in February 1917. The accompanying War Diary entry reports that the enemy blew a mine forming a 100ft-diameter crater, probably in response to '... our repair work in 6ZA. We lost 2 men killed and wounded.'

fired by 171 Tunnelling Company in the Ypres Salient on 17 February 1915.

Two months later mines that they had prepared formed part of a coordinated attack on Hill 60 in the same area. Soon offensive mining became a regular part of most major operations, the assault on the heavily fortified Messines ridge in June 1917 being the prime example. Day by day across the front the most incredible bravery was shown by men working underground in imminent danger of death from enemy counter-mining, carbon monoxide poisoning and subterranean battles.

By the time of the Somme battle a third role for the tunnellers was in demand, the preparation of approach tunnels to forward gun positions and OPs. In addition 'Russian Saps', shallow tunnels leading forward to be broken out of by the assaulting troops at zero hour, were successfully used. In the battle of Arras (April 1917) use was able to be made of the extensive cave system under the town, sufficient to house 11,000 infantry and quantities of stores. From there some three miles of subway were driven towards the front line to provide a covered approach for the assault.

As the war evolved the tunnellers' skills became increasingly in demand for dug-out command posts, signal and dressing-stations, magazines and stores and troop accommodation. Their familiarity with underground workings then enabled them to tackle, by means of specially trained teams, the proliferation of delayed action mines, booby traps and similar devices that were encountered during the final advance in 1918. At this stage the units, with their wealth of trade skills and transport, proved particularly useful in the restoration above ground of the chaotic communications so vital to maintaining the forward momentum of the advancing armies.

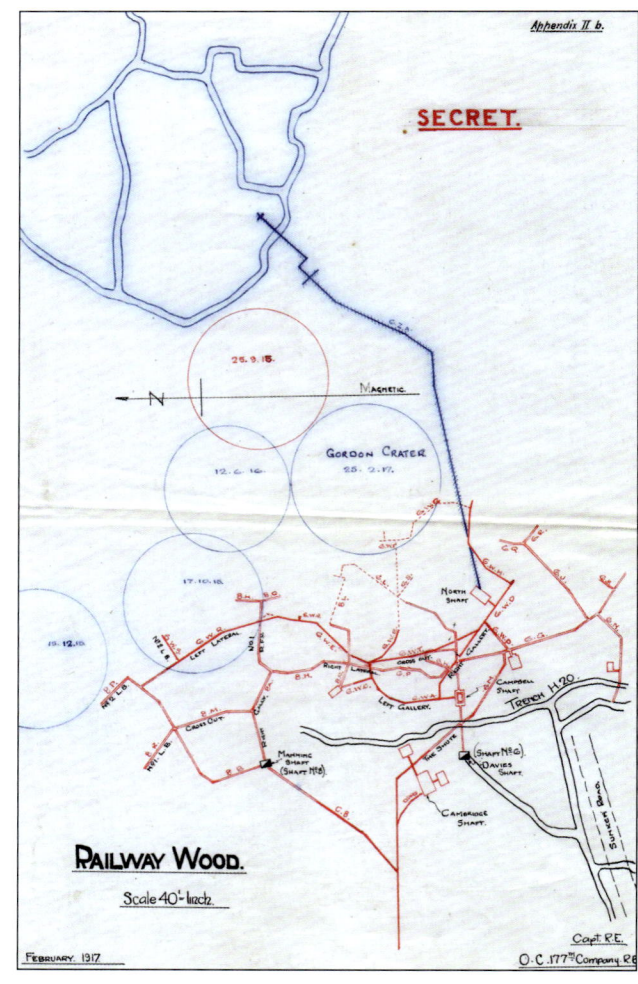

As the ramifications of the war multiplied, so did the variety of the units that had to be created to cater for them. The Chief Engineer's Christmas card of 1917 (see page 131) makes the point. The two major specialisations outside the Engineer-in-Chief's responsibilities were Transportation ('RE Tn', as it was known), and Signals, which have their own chapters in this book 'Movers and Shakers' (page 184) and 'Wire and Wireless' (page 115). (See also pages 106 and 117 for the Corps role in the introduction of tanks and the birth of the Royal Flying Corps.) By the end of the war, of the 315,000 officers and men in the Corps, 85,000 were in Transportation. Of 1,330 RE units (excluding Transportation) in the order of battle in 1918, 589 were Signals, although it only includes units under his direct command.

The main specialisations other than Signals and Transportation are the subject of this chapter.

GAS

A month after the shock of the first German use of gas at Second Ypres in April 1915, the decision was made to retaliate in kind. The Corps was the obvious choice to take on the challenge of '… dealing with this gas question as a whole', as the Chief of Staff at GHQ expressed it. Major Charles Foulkes was charged with developing the means and the organisation to provide an offensive capability in the field. This first bore fruit in September 1915 at the battle of Loos in the form of cloud discharges of chlorine from 5,500 cylinders distributed across a 24-mile front in sectors totalling about 14,500 yards. It was a stumbling start. The wind was less than ideal, but the Corps Commander, Sir Douglas Haig, stuck to his plan, which was predicated on the use of gas. In some areas the discharges had to be curtailed from local commanders' cylinders to avoid casualties to

Below right: Major General C. H. Foulkes CB, CMG, DSO, who in 1915, while still commanding 11 Field Company, was summoned to create the British offensive gas capability for the Western Front. Foulkes finished the war as Director of Gas Operations. He had had an exceptionally adventurous early career (see also page 90). His DSO had been awarded for gallantry in the first battle of Ypres. Foulkes remained in the chemical warfare business throughout the war, becoming Director of Gas Services in 1917. After the war he saw service in India and Ireland and his last appointment on the active list was as Chief Engineer Aldershot Command.

Far right: Sergeant Martin Fox handling gas cylinders in the forward line.

Right: Defensive measures against gas were originally an RAMC responsibility but the Corps took them over in 1916. The first respirators were simple flannelette bags, as seen here, impregnated with a suitable chemical. Later the box respirator came into service, which allowed for appropriate filters to be added as new gases were introduced.

Above: The Livens Projector was the brain-child of Captain W. H. Livens, a company commander in the Special Brigade. It comprised a tube capable of firing a 30lb phosgene bomb, or equivalent alternative warhead. The tubes were set in the ground as shown and fired electrically.

Right: Corporal James Dawson, VC, whose 'most conspicuous bravery and devotion to duty' during a gas attack on 13 October 1915 earned him the Victoria Cross. Gas started leaking as the infantry were forming up. Dawson, under intense enemy fire, moved the cylinders forward out of the trench and fired rifle bullets into them to allow the gas to escape.

their own troops; in others, however, they were successfully emptied. The lessons were learned and the morale effect on the enemy made it a worthwhile start.

In 1916 the 'Special Brigade RE' (a deliberately covert title), was formed with twenty-five companies, including sixteen cylinder companies in four battalions, four mortar companies in one battalion and Z Company employing flame projectors, to manage what was to become an indispensable part of all operations. The 4-inch Stokes mortar (see page 132) proved an effective launcher. At Loos it was used only with some improvised smoke bombs; these then caught on with the infantry and so began the use of smoke as cover for infantry movement. The 3-inch version of the Stokes mortar also became a success, tens of thousands being produced. The use of the Livens projector (see illustration) was also much exploited. As many as 4,000 of these projectors could be fired simultaneously in a single discharge using an electrical device. They were first used at the opening of the battle of Arras in April 1917.

CAMOUFLAGE

The French led the way in recruiting artists into the business of camouflage, primarily to protect gun positions from observation. For the British Army, the artist, Mr Solomon J. Solomon, RA, went to France with a team of artists and stage scenery experts in 1915 to advise, and by March the following year the 'Special Works Park RE' was in business. Solomon was commissioned into the Corps as a lieutenant colonel (thus acquiring the rare post-nominal letters 'RE, RA') and many talented artists joined him.

The camouflage and deception business burgeoned. Two factories were set up, at Amiens and Aire. Their main products were screens for gun positions and observation posts in ever more crafty disguises, for example as trees and even corpses. But their ingenuity was greatly in demand in a wide variety of activities, such as in the obligatory concealment of spoil from tunnelling operations, and their work saved many hundreds of lives. Nor were they themselves always out of danger as testified by one DSO, five MCs and one DCM.

Above: Searchlights against the Zeppelins, an early depiction on a postcard.

SEARCHLIGHTS

Searchlights, in service originally for coastal defence, had been retained in the Corps for possible battlefield illumination. The Officer Commanding the School of Electric Lighting at Gosport viewed with some scepticism the idea of lights illuminating no-man's-land in France and quietly proceeded to design a U-shaped mounting that might serve to elevate his 60 cm lights in the anti-aircraft role. Thus he was

Below: War Babies, the 1917 Christmas card produced by the staff of the Engineer-in-Chief's headquarters to illustrate the number and variety of the units under his command. Represented are: Searchlights, RE Park, Base Park Company, Tunnelling Company, Artisan Works Company, Workshops Company, E&M Company, Railway Company, Corps Troops RE, Boring Section, Forestry Company, Land Drainage Company, Sound Ranging Company, Pontoon Park, Road Construction Company, Port Construction Company, Inland Water Transport, Special Companies (Gas), Siege Company, Army Troops Company, Special Works Park (Camouflage), Signals.

XMAS 1917

WAR BABIES.

ready in the autumn of 1915 to respond when the increasing Zeppelin raids had driven AA Command to ask for his services.* The Tyne Electrical Engineers and London Electrical Engineers carried most of the expertise and provided much of the manpower in the early days.

Searchlights were immediately successful. The Zeppelins were driven higher and were countered by 'aeroplane barrages' supported by lights. The air defences around London and other key target areas grew. Forty-two anti-aircraft companies RE were established, stationed around the country. Daylight raids by aircraft were a menace for the first eight months of

*The Admiralty was responsible for anti-aircraft defence in the early part of the war in view of the demands on the Army from France. Most of the guns initially were naval 13-pounders. As the system expanded the Royal Artillery played an increasing role and a gunner major-general took command of the London Air Defence Area in 1917.

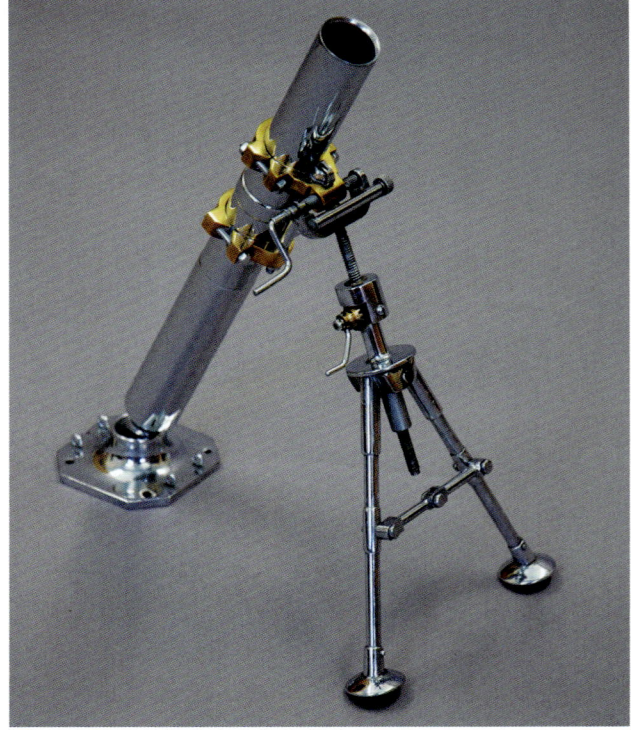

Left: Trench warfare equipment was produced both in the RE Workshops in Flanders and, later in England, by the Trench Warfare Department under Brigadier General Louis Jackson who came back from retirement to undertake this task and ended the war as Major General Sir Louis, KBE, CB, CMG. He and his staff had to sort out countless bright and not so bright ideas that flowed in both from the front and the general public and turn them into practical products. The Mills bomb and the Stokes mortar (model illustrated) were both highly successful examples, taking their name from the responsible members of Jackson's staff.

Above: Major Peter Nissen was commissioned into the Corps in 1915 at the age of 41. He came from an engineering family who had emigrated to the United States from Norway in the 1850s. He qualified as an engineer in Canada, and practised there, in America and South Africa, before coming to England in 1912. Nissen foresaw the need for hutting early in 1916, conceived the idea of the famous semi-circular shape with its simplicity of manufacture, transportation and construction, and was prudent enough to apply for a patent. One hundred thousand of the standard 27-foot hut were sent to France and a further 10,000 of the 60-foot hospital version. The hut and its many variants became Nissen's livelihood after the war.

1917. Thereafter the Germans took to night attacks, using both aeroplanes and airships. The last attack on London was in January 1918.

Across the Channel, a calamitous German raid in June 1916 on a base ammunition depot that blew up 40,000 tons with the last bomb dropped by the last aircraft in the sortie, galvanised GHQ into providing for a rapid build-up of men and equipment. Up to that time only modest arrangements for air defence had been made the previous year. By the end of the war the searchlight force in the BEF had built up to more than 2,000, comprising forty-five sections equipped with new 90cm projectors replacing the earlier 60cm versions.

ENGINEER SUPPORT

From the UK through to the front the myriad necessities of life had to be met. To cope with the vast expansion of the Army in 1914, accommodation had to be found almost immediately for 800,000 men. The immediate inflow was catered for by squeezing existing barracks, and arranging for hirings, billets and tented camps.

Soon more permanent hutted camps were needed. A sectional hut was designed in the War Office, new camp sites selected and contracts let, including for hospitals, training areas, horse lines and POW camps.

In the war zone the accommodation problem did not become urgent until the build-up for the Somme. Then tents, tarpaulin-covered frameworks and improvised shelters were used, grouped around villages and farms. By the end of that battle it became clear that something more permanent would be needed. So arose the idea of the Nissen hut (see illustration), the first of which appeared, as part of the initial order of 47,600, early in 1917, benefiting the preparations for the battle of Arras.

A phenomenal base area back-up grew to supply these and the many other needs of the Army in the field, too complex to be covered here but embracing what are nowadays known as Resources. Some statistics must suffice: see table overleaf. This scale of engineer support was, however, no more than necessary for an

Below: The Duckwalk, another watercolour by Captain Richard Cooper. The materials for these walkways and trench duckboards were produced in prodigious quantities by the RE workshops and some by local contract. At the height of the Passchendaele battle, when the mud was at its worst, 375,000 trench boards were sent to the front, enough to make 400 miles of pathway.

enterprise catering for the needs of more than fifty divisions at the front and all the logistics stretching back to the Channel ports. The Engineer-in-Chief (see page 135) was responsible overall through his corps and army chief engineers and his Director of Works. He had come a long way since 1914 when his gunner opposite number at GHQ BEF, had remarked 'I don't suppose you will have much to do in this war.'*

*Signals and Transportation had their own directors at GHQ.

Daily tonnage of all types of stores to be moved (estimate for 1917)	40,200
Daily tonnage of engineer stores to be moved	18,400
Daily supply of stone (some brought in from the Channel Is)	3,250
One month's imports of stores from England	33,560
In November 1917 the Works Directorate contained 171 officers, 540 subordinate staff, three base park companies, eight stores sections and thirty-one other RE units	
Thirteen officers, five RE companies and three and a half labour battalions were employed on forestry	

Left: A sawmill being set up on the Amiens–Albert road in November 1916. The huge demand for timber resulted in the formation of eleven special forestry companies under their own directorate. (IWM Q4610)

Left: Light railway wagons loaded with heavy shells, Elverdinghe, near Ypres, February 1917. Trench 'tramways', often man- or horse-drawn, were set up in forward areas as early as 1915. More formal narrow-gauge systems, such as that illustrated, (powered by a petrol-engined Simplex locomotive), did not develop widely until 1916. (IWM Q1696)

25 | THE HUNDRED DAYS

MARCH TO MAY 1918

Among the Allies, a sense of frustration prevailed at the end of 1917 (see page 125). The collapse of Russia, no breakthrough on the Western Front, and mounting casualties from Passchendaele and Cambrai brought disillusionment in Britain. The British Prime Minister, Lloyd George, lost confidence in Field Marshal Haig and withheld reinforcements to his armies. This was offset by improved unity of command in France (under Marshal Foch) and the build-up of American forces since the United States' declaration of war in April 1917. However the American deployment was slow; by May 1918 they had fewer than half a million men (with only one division in the line) compared with over two million from both the British and French.*

The Germans then launched three mighty separate offensives: in March, towards Amiens, in April in Artois towards the Channel ports, and in May, in Champagne across the River Aisne to the west of Paris. All fronts gave way after desperate fighting against new German tactics, sappers acting as infantry in most of the battles. Paris came within range of German artillery. Haig issued his famous 'backs to the wall' Order of

*The number of divisions that participated in the final advance were: British 54, French 42, United States 17.

Left: Lieutenant General Sir George Fowke KCB, KCMG was the first Engineer-in-Chief on the Western front in the 1914–18 war. In August 1914 he had deployed to France as Brigadier General Royal Engineers. At that time this was the senior engineer appointment – and a purely advisory one – at GHQ. His staff consisted of one warrant officer clerk (bicycle-mounted) and his transport was one motor-car which he shared with his opposite number the Brigadier General Royal Artillery. From this comparatively lowly position, Fowke presided over an expansion of the Corps and its activities which remains unequalled in its history and which was epitomised in the famous cartoon 'War Babies' (page 131). In 1915 Fowke was promoted major general and his appointment upgraded to that of Engineer-in-Chief. Later, in the rank of lieutenant general, he became Adjutant General to the British Expeditionary Force.

Below: The St Quentin Canal, the crossing of which initiated the breaking of the Hindenburg Line in September 1918. In this picture men of the 46th Division, some still wearing their lifejackets (see text) are being addressed by their brigade commander, Brigadier J. V. Campbell VC. (IWM Q9535)

BRIDGES IN THE 1914–18 WAR

On the outbreak of war, the only bridging equipment in the Corps consisted of timber framed and clad pontoons, steel 'Weldon' trestles and some light rafting equipment based on folding boats. The latter were held in the field companies, which also carried two pontoons and a trestle. Two bridging trains carried a back-up of forty-two pontoons with fifteen feet of superstructure and sixteen trestles. With some modifications these equipments all did excellent service throughout the war albeit in vastly increased quantity.

To deal with the problem of the civilian road and rail bridges having fallen victim to the exigencies of war, 'Stock Span' bridges were produced to replace them at standard load capacities and spans. These were subsequently upgraded to meet the needs of heavier artillery, motor transport and the arrival of the tank.

Quicker builds were achieved by the introduction of the Inglis bridge, developed from the pre-war design of an infantry bridge by Charles Inglis, a fellow of King's College, Cambridge. This was the first equipment dry bridge to enter service. The Inglis bridge was made from tubular steel

Right: A 180-foot Hopkins bridge crossing the Canal du Nord in the Third Army Cambrai sector, as depicted by Major Tom Alban in 1919. The New Zealand Tunnelling Company with 565 and 577 Army Troops RE built the bridge in September 1918 in support of the final advance. In this phase of the war most of the tunnelling companies had to turn their hand to other work. Some became particularly adept at clearing the many booby traps that the retreating Germans left behind.

Right: A Sankey bridge, an up-grade of the standard pontoon bridge to take heavier loads, using steel road-bearers and close-spaced pontoons. In the background is a Pont Levis bridge, a special design to allow bridges to be lifted for river or canal traffic.

girders and could be quickly erected. Originally of triangular cross-section, later versions were developed in the more conventional rectangular through-bridge form with overhead bracing. Towards the end of the war the Stock Spans were replaced by the Hopkins bridge.

The final advance in 1918 produced a colossal demand for bridging, for example, almost twice the number of standard-span heavy bridges than had been built in the previous three years of the war. In addition to these, and light bridges possibly running into thousands, more than 200 were built using salvaged materials, as illustrated.

Outside Europe, all theatres required the same sort of field-bridging operations as experienced in Europe. In Egypt the Suez Canal presented a special problem, eventually solved by timber-piled piers connected to floating landing bays with a lifting arrangement to allow shipping to pass. In Palestine, the Jordan was crossed using pontoons, eventually replaced by the famous three-span steel girder Allenby bridge. In Mesopotamia the pontoon trains became critical to many operations and the Tigris was crossed seventeen times by the Bengal Sappers.

Left: An 84-foot span Inglis bridge Mark I at La Motte, April 1918. In the foreground is a Mark II Stock Span bridge.

Left: Bridge 4 Essex Farm (across the Yser Canal north of Ypres), a watercolour by Captain W. F. C. Holden (Royal Scots), a 'camouflage artist'. The bridge was crossed by tanks during the Third battle of Ypres (Passchendaele).

the Day. However, in the north, Ypres remained solid. The Americans stood firm in the south. The attacks lost momentum and lacked any reserves for exploitation. In the German High Command, General Ludendorff had to abandon his master plan of a breakthrough in Flanders that had depended on the success of the offensives.

AUGUST TO NOVEMBER 1918

Even before the May offensive the Allies had been planning their own move and this resolved itself into the decisive 8 August battle of Amiens in which a combined assault was made by the British Fourth and French First Armies under Haig's command. Total surprise was achieved as the troops advanced into the morning mist supported by more than 400 tanks and 1,400 aircraft. The Germans lost more than 70,000 men and all Allied objectives were achieved. Most importantly, the German High Command had to accept, on what Ludendorff called the 'black day' of the German Army, that the morale of their soldiers was no longer up to maintaining the fight.

In front of the Allies now lay the Hindenburg Line, the 90-mile labyrinth of concrete and wire to which the Germans had withdrawn early in 1917. Integral to it were the Canal du Nord and the St Quentin Canal. This formidable obstacle was broken at the end of September at the start of the main Allied counter-offensive that was to continue almost without pause until the end of the war.

For the sappers, this phase of the war meant a sudden change in attitude from the defensive to the offensive. Bridging was in urgent demand, roads had to be repaired, routes cleared, bombs and booby traps disposed of and water found and brought up to the leading units. The means by which the infantry could cross the canals had to be improvised, using cork or petrol-can floats, folding boats and canvas stuffed with hay. When the 46th Division (CRE Lieutenant Colonel H. T. Morshead)* made their epic assault on the St Quentin Canal at Bellenglise on 29 September, many of the infantry also swam

*Henry Morshead, surveyor, explorer and mountaineer. He was a member of the 1922 second Everest expedition, failing at about 25,000 feet from exhaustion and frostbite. In 1927, now lacking the tops of three fingers, he took part in the first crossing of Edge Island, Spitzbergen. He was murdered in Burma in 1931.

Left: Sikh sappers blasting an artillery road over the 'Ladder of Tyre', pencil and watercolour by James McBey. This narrow cliff path with gradients of 1 in 5 lay on the route of 7th (Meerut) Indian Division in their advance from Acre towards Beirut. Three days of blasting and clearing the rubble were needed to make the road passable. (IWM PIC 2392)

Left: German Machine-Gun Pit, Aveloy, a watercolour by Sapper Bert Wardle, one of the 'camouflage artists'.

across wearing lifejackets, 3,000 of which were requisitioned from the Channel ferries. Ladders had to be provided with which to scale the almost perpendicular banks of the canal cutting. The success of this division, by such meticulous planning and dogged determination, triggered a breakthrough along the whole line where less progress had been made initially. It was 5 October before the Hindenburg Line was fully penetrated across the whole front.

The Germans now sought an armistice. Elsewhere there were manifest signs of collapse in the will to fight. On the day of the offensive against the Hindenburg Line the Bulgarians succumbed in Salonika. The previous week, Allenby had broken through in Palestine and the Turks were in retreat. The Italians, who had defeated an Austrian offensive on the Piave in June, were preparing for a counter-offensive in October.

Despite all this, the Germans fought hard to secure the best position they could in the Armistice negotiations. Some of the fiercest fighting took place around Cambrai where the Canal de la Sensée and Canal d'Escaut provided strong

check lines. The last major battle was fought twenty miles farther east on the Sambre–Oise Canal when First, Third and Fourth Armies attacked this line. Strong resistance was met in the sector around Ors and Catillon. In quintessential sapper operations three VCs were won, by Major Arnold Waters and Sapper Adam Archibald north of Ors and by Major George Findlay at Lock One south of Catillon. Two days later, the last VC of the whole war was also won by a sapper, Major Brett Cloutman (see also page 127), for carrying out a solo operation to cut the leads on a bridge prepared for demolition.

The Hundred Days had been victorious and the finest achievement in British military history to date. It had cost about 290,000 casualties on top of the 250,000 earlier in the year. Overall in the war, the Corps had expanded to a greater size and breadth of responsibility than at any time in its history. Although the Armistice brought to an end the war in Europe, there was little rest for the Corps as it became involved in campaigns in Russia, the Middle East, Afghanistan and India (page 141).

Left: Lock 1 on the Sambre–Oise Canal photographed in 1997.

Right: Major George de Cardonnel Findlay, VC, MC, who was awarded the VC for his resolution and outstanding bravery in the crossing at Lock 1 of the Sambre–Oise Canal on 4 November 1918. His field company suffered many casualties in the heavy bombardment that temporarily stopped the attack. He himself was wounded and much of the equipment damaged, but he pressed on and was the first man across the Lock in the assault by the 2nd Battalion the Royal Sussex Regiment. Findlay had earlier won the MC during Third Ypres (Passchendaele), for working '… for hours in heavily shelled areas and deep mud to establish communications'. He later wrote of that period: 'Yet in the midst of all this danger and desolation, there is a certain fascination out there on patrol …'

Above: The medals of Major (later Colonel Sir) Arnold Waters, VC, DSO, MC, and Sapper Adam Archibald, VC (see also text). A later account reads: 'Only through the heroism of Major Waters and Sapper Archibald of the 218th Field Company was it possible to get a bridge across. The whole area was swept with shell and machine-gun fire, and it seemed impossible for anyone to live on the bank of the canal. All the rest of the party were killed or disabled, yet these two gallant engineers carried on the work, while bullets splintered the wood they were holding and struck sparks from the wire binding the floats.'

Right: The Royal Engineer War Memorial was unveiled at Brompton Barracks on 19 July 1922 by HRH The Duke of Connaught. It was designed by Messrs Hutton and Taylor, both FRIBA, and sculpted by Mr Alexander Proudfoot. The most distinguished of the 19,800 sappers who were killed or died on active service, Field Marshal Earl Kitchener, is the only name on the 70-foot obelisk, inscribed simply as 'Kitchener, 1850–1916'. A large sum was also invested to provide help for the education of the children of officers and soldiers of the Corps. After the Second World War the decision was taken not to erect a further memorial to the 10,800 who died, but to add '1939–1945' to the same memorial and invest the money raised in homes for the families of members of the Corps in need of help.

26 | IMPERFECT PEACE

The 1919 Treaty of Versailles left much unfinished business outside Europe. Russia was in turmoil and five field companies found themselves there in support of the British mission in the Caucasus, together with a sorely needed transportation organisation, until October 1919. The Irish rebellion, blowing up in earnest in 1918, absorbed much effort until Southern Ireland was granted dominion status in 1921. But the main operational business of the Army was imperial policing in the Middle East and India.*

*See page 65 for the Third Afghan War (1919–20).

Above: Lieutenant General Sir John Glubb, KCB, CMG, DSO, OBE, MC ('Glubb Pasha') (1897–1986) shown in the uniform and head-dress of the Jordanian Army (The Arab Legion), which he commanded from 1939 to 1956. He served with 7 Field Company in the 1914–18 war, was thrice wounded and awarded the MC. From 1920 he was seconded to the Colonial Office as political officer in Iraq. In 1926 he retired from the British service to take up a similar appointment under the newly formed government of Iraq. Glubb was successful in raising a police force recruited from the Bedouin to pacify an area in which local feuding had become a way of life. In 1930 he went to Transjordan, on the invitation of Emir Abdullah, to set up a similar system of policing the border area. His success there led to his appointment as commander of the Arab Legion. He led them in their only campaign of the Second World War, following the 1941 German-inspired revolt in Iraq, and remained in command until political developments in the Middle East forced his retirement in 1956. At his memorial service, King Hussein, who attended in person, described him as '… a down to earth soldier, with a heart, a simple style of life, an impeccable integrity, who performed quietly and unassumingly, the duties entrusted to him by his second country, Jordan, at a crucial moment in its history …' **Left:** His medals as worn.

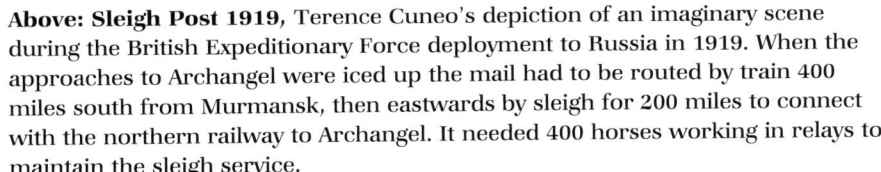

Above: Sleigh Post 1919, Terence Cuneo's depiction of an imaginary scene during the British Expeditionary Force deployment to Russia in 1919. When the approaches to Archangel were iced up the mail had to be routed by train 400 miles south from Murmansk, then eastwards by sleigh for 200 miles to connect with the northern railway to Archangel. It needed 400 horses working in relays to maintain the sleigh service.

Most of this commitment was met by proxy through locally raised forces and, in the case of India, by the Indian Army whose engineer units were officered by the Corps. After 1919, no Royal Engineer units were deployed overseas on operations although many were stationed round the globe in peacetime locations.

During the inter-war years, the Royal Engineers suffered with the rest of the Army the starvation of the means to modernise. Nevertheless, the innovative spirit was alive and well and perhaps of all arms the Corps was the least devoted to the horse. Major Giffard Martel was busy at Christchurch (see page 118) and mechanisation was the spirit of the times, inspired by the exigencies of war. By 1918 not only had the searchlights, pontoons and E&M companies become vehicle-mounted, but three completely mechanised RE battalions had been set up for operating with tanks. These promising moves were stifled by economies until the late 1920s when 17 Field Company, under Martel, became the trials unit for all the necessary equipment, and by 1937 all field companies were mechanised.

Above: The Albert Medal was issued either in gold (1st Class) or in bronze (2nd Class). In the centre is the monogram 'VA' surrounded by the legend 'FOR GALLANTRY IN SAVING LIFE ON LAND'. It was instituted in 1877. (An earlier version was instituted in 1866 for saving life at sea.) Known generally as 'The Civilian VC', the gold version was abolished in 1949 in favour of the George Cross and earlier recipients were invited to exchange their Albert Medals for the GC. An example was that awarded to Lieutenant (later Lieutenant General Sir) John Cowley for his bravery in rescue work following the 1935 Quetta earthquake.

Above: Experimental tanks at Christchurch in the 1920s.

Left: A motor tool cart, a visionary idea produced by an East Anglian field company in 1911 for a possible mechanised field company. Based on a 40hp Daimler, two oak box-girders gave it a gap-crossing capability.

Gradual advances were made, too, in the fields of bridging (see page 161), demolitions and mines, searchlights, earth-moving plant and water-supply equipment. In general, Army life in the inter-war years had more professional activity to offer the sapper (officer and soldier) than his infantry or cavalry opposite number; not least because the continuing commitment to provide the Works Services of the Army meant enforced periods of employment in that unglamorous field. Generally despised, 'works' appointments could lead to some excellent postings, such as on the construction of the naval base at Singapore.

Right: The Waziristan Circular Road was a major project for the Indian Engineers during the period 1922–4. Its primary aim was to support the fortified camp at Razmak, 7,000 feet above sea level, to allow for the rapid deployment of troops and to encourage trade in the North-West Frontier area.

Below right: Protection on the road.

Below: An Inglis bridge on the North-West Frontier.

Left: The Waziristan Campaign Fort commemorates operations on the North-West Frontier in the years between the wars. It is a replica of a typical hill fort, complete with rickety ladder, such as were the demolition targets for many frontier operations. The piece was a gift from the Pakistan Engineers on their formation in 1947.

Left: The Indian General Service Medal 1908–1935 with clasp 'Afghanistan NWF 1919' and Mentioned in Despatches oakleaf, awarded for the Third Afghan War (1919–20).

FIRST SHOTS

The four-division BEF that deployed to France in October 1939 bore little resemblance to its namesake of 1914. It had been put together at short notice, under-equipped, under-trained and with no clear objective. However, seven months of 'Phoney War' were to elapse before it was called upon to fight. The first shots at the Germans in a serious manner were to be fired far to the north in the confused efforts to oppose the invasion of Norway. The Allies' (British, Norwegian, Polish and French) first objectives were to seize Trondheim and Narvik.

On 18 April 1940, 55 Field Company under Major Sir John Forbes landed with the Trondheim force to become the first sappers in action in the war. Two weeks later, this force had to withdraw in the face of the overpowering German response and total air superiority. Farther north, the battle for Narvik, including a hard-fought but expensive diversion farther down the coast, was on a larger scale. It began with a successful naval attack

Above: General Sir Frank Simpson GBE, KCB, DSO (1899–1986) was Director of Operations and later Vice Chief of the Imperial General Staff during the 1939–45 War. He had been commissioned in 1916 and went to India in 1919 to serve with the Bombay Sappers and Miners. Thus he was a veteran of the 1914–18 and Third Afghan Wars when he went to France with the BEF in 1939. There he was awarded the DSO for his part in the defence of Arras. However, his potential had already been spotted by the future Field Marshal Montgomery, under whom he served in several appointments. He was GOC-in-C Western command from 1948 to 1950 and Chief Royal Engineer from 1961 t0 1967.

Below: The demolition of a bridge in Louvain, a photograph taken during the BEF retreat from the line of the River Dyle in 1940.

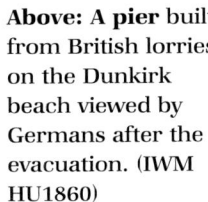

Above: A pier built from British lorries on the Dunkirk beach viewed by Germans after the evacuation. (IWM HU1860)

Left: 691 (Mowlem) General Construction Company, raised by John Mowlem and Company, working on a military airfield in France in 1940–1. By the time of the German invasion, four concrete runways had been built and nearly thirty new airfields levelled and sown with grass by such companies. (IWM F4888)

and ended with a flourish when the Germans were forced to abandon the port and destroy its facilities. But the realisation that Britain was in no position to cling even to this small toehold of mainland Europe led to the evacuation of the whole force on 7 June, three days after the last man had left the Dunkirk beaches.

BEF

The Phoney War came to a violent end on 10 May 1940. Eighty-nine German divisions, ten of them panzers, divided into three army groups, burst into Belgium and France on three main fronts. The British, deployed along the River Dyle to the east of Brussels, faced the northernmost of these. By that time the teeth of the BEF had grown to thirteen divisions of which five were in the front line in I and II Corps, two in III Corps farther back on the Escaut and two in general reserve. Of the remaining four, three were still in training in-theatre and one, 51st Highland Division, was detached under French command.

Each division had its three field companies (normally one regular and two TA) and a field park. However, in the early days of 1940 these were 'raided' to provide a centralised force (X Force) for general works, principally to construct the lines of pill-boxes and tank obstacles deemed necessary to support the defensive battle that was foreseen. In due course these companies returned to their divisions as reinforcements arrived, not least those specially raised from the engineering industry, some civilian firms even producing their own units virtually complete.

Above: Pill-boxes, supplemented by dummies as illustrated, were constructed in 1940 all over Britain to form defence lines, primarily anti-tank, against the expected German invasion. This mammoth task, undertaken by contractors and Royal Engineer units under the auspices of Royal Engineer Works Services, absorbed a major proportion of the country's resources of construction material during 1940. This static defence policy was revoked when General Sir Alan Brooke returned from France and replaced General Ironside. (IWM H4847)

Of the other critical engineering tasks, airfield construction and maintenance, a wartime responsibility of the Corps, was a high priority, including the ancillary buildings. Survey and Transportation (see pages 176 and 184) were also major commitments.

The four-week fighting withdrawal from 10 May to the evacuation from Dunkirk, and the eventual fate of those whose route to safety lay outside the encircled BEF can be simply described. The Dyle line was abandoned on 19 May and a new line along the Escaut taken up. Special groupings had to be made, notably at Arras, to protect the southern flank of the BEF as the German Central Army Group A broke through towards the Channel ports, reaching Amiens on 20 May. Boulogne fell on 25 May. The Belgians, under immense pressure for two weeks, capitulated on 26 May, leaving an alarming gap on the British left. Evacuation from Dunkirk was ordered on the same day. Within its perimeter the enemy had to be held back and the embarkment organised. By 4 June nearly 225,000 men had been saved. Another 140,000 remained in France and were being reinforced on the assumption that the fight would continue. However, the whole of the French IX Corps was surrounded on the coast at St-Valéry and, with 51st Highland Division, was forced to surrender. In mid-June the French sought an armistice and as many as possible of the British men and their weapons and equipment were evacuated through Nantes, St-Nazaire, Brest and Cherbourg.

These facts conceal what had been achieved by an inadequately prepared and largely citizen army in its fight against the German onslaught. From 10 May onwards it had been a hand-to-mouth existence in which countless feats of gallantry were performed and ingenious expedients resorted to in a bid to stem the tide. Aside from their own normal job of firing demolitions and getting the formations back across the refugee-cluttered roads, most sapper units fought as infantry at one time or another as the gaps in the line had to be filled. Many of these were the very lines-of-communication units whose individual members had not even contemplated putting on uniform a year earlier. 13 Field Survey Company proved their field engineering skills by successfully destroying the bridges over the Yser to close a gap after the Belgians had withdrawn. A chemical warfare company found the Germans already in occupation of one of their bridges, forced them off and drove a lorry loaded with explosives on to the bridge as a rapid demolition device. Three companies of the 4th Division Engineers (7, 59 and 225) were ordered to hold a position on the Lys–Quesnoy Canal, the bridge over which came under heavy machine-gun fire, so threatening the division's withdrawal. The sappers mounted a counter-attack with two tanks of the 13th/18th Hussars and a platoon of the Black Watch and so held off the enemy out of range of the crucial bridge.

By such means, underwritten by its sapper support, the BEF had saved itself to fight another day. It remained for the Royal Navy to recover it to Britain. Within the Dunkirk perimeter more than 600 bridges were demolished and countless craters blown. On the beaches improvised jetties were built and the sappers' watermanship skills were in full demand as the survivors were gradually embarked albeit amid the mayhem and

Below: St Paul's Cathedral surrounded by the havoc of the Blitz. The George Cross awarded to Lieutenant Robert Davies who, with Sapper George Wylie, dealt with a deep buried bomb there, is on display in the cathedral. **Right: The bomb disposal** team working on the St Paul's bomb.

mounting casualties as the Luftwaffe did their best to frustrate the operation.

HOME FRONT

As much as eighteen months before the war the Corps had been fully engaged in works to improve the defences of the UK and to cater for the anticipated expansion of the Services. Anti-aircraft defence was high on the agenda; by April 1940, the numbers employed on the air defence of Great Britain were 154,000, about 50 per cent more than the whole peace garrison of the British Isles. This included twenty-one infantry TA battalions converted to Royal Engineers in the searchlight role together with four sapper units. The Corps had provided the searchlight units until 1938 when the transfer of the service to the Royal Artillery was approved, to be implemented in a gradual process that had been effected by 1941 (see also page 148). The vast programme of works for such as ordnance depots, barrack

Above: The Defence Medal was awarded for service in Great Britain or overseas territories subjected to enemy action. It has a reverse showing the Crown resting on the stump of an oak tree, flanked by two lions with the dates 1939 and 1945.

accommodation and training camps, at a time when offices had been denuded by the demands from across the Channel, brought hundreds of reservist and retired officers back to the Colours.*

On top of all this there was now, for the first time since the Armada, a serious possibility of invasion. Britain became a theatre of war. Pill-boxes sprung up around the country. Coastal defences, including elaborate minefields, were prepared. Fortunately 1st Canadian and 2nd New Zealand Divisions had arrived and, with 1st (British) Armoured Division, formed VII Corps for the defence of the UK. The Canadian contingent, which included Corps troops, was particularly strong in sappers and engineering equipment. A second Canadian division arrived at the end of 1940.

Airfield protection called for special measures. The first bomb attack was on 5 June 1940. By the end of that month no less than 134 airfields had RE maintenance parties in position, many provided by the construction companies that had returned from France. The intensity of the raids grew. Manston, for example, suffered one every day in August, September and October during which period the air defences of Britain, including searchlights and artillery, were at full stretch. The airfield maintenance system therefore played a crucial part in winning the Battle of Britain.

BOMB DISPOSAL

Dealing with unexploded bombs now became an urgent matter. Some special bomb disposal parties had been formed by the end of 1939, but the imminence of air attack brought matters to a head in May 1940 with the raising of 220 teams by July. These then, with the personal interest of the Prime Minister, were formed into a complete organisation under Major General G. B. O. Taylor. The outstanding courage of these men, working in terrible conditions, often deep underground, mastering the ever-increasing complications of the German bombs, ranks as one of the proudest achievements of the Corps. The casualty figures of 394 killed and 209 wounded bear eloquent witness to the hazards that they faced (see also page 183).

*One garrison engineer was 82 years old.

OVERSEAS

By the spring of 1941 the threat of invasion had receded and the UK was established on a war footing as a base for future operations in Europe. Overseas, however, there was little to bring good cheer. In the Mediterranean and Africa, some fine successes against the Italians in 1941 were offset by German intervention in North Africa and Greece, so forcing the British and Commonwealth forces into stop-gap strategies that could only be remedied by the gradual build-up of men and equipment. In the Far East, 1941 saw the efficient and ruthless Japanese bid to usurp the European colonial regional power. Hong Kong fell in December 1941 and Singapore in the following February. By then Burma had already been invaded and, after a resolute fighting withdrawal, the British were back in India by mid-May. These events and their repercussions are covered in Chapter 30.

Right: Summer at Colditz – the British Corner. Colditz (Oflag IVC) became the final home of many sapper prisoners-of-war, whose skills and experience made them potential escapers. Major (later Brigadier) Montagu Champion-Jones, who painted this watercolour, was moved there after a spell at Laufen. He had been captured in 1940 during the Blitzkrieg after six days on the run after orders had been given for small parties to attempt to break through to the Channel ports.

Above: A Box Girder bridge built across the bomb crater at Bank tube station in January 1941 by 691 General Construction Company to restore traffic flow from Cornhill to Queen Victoria Street and allow the reconstruction of the station to proceed. The bomb exploded in the booking hall and resulted in the largest crater ever caused by a bomb up to that time.

Left: The 90cm carbon arc searchlight was the workhorse of the searchlight units in the 1939–45 War. The inset shows the equipment in action with the Tyne Electrical Engineers.

Above: Colonel Stuart ('Archie') Archer, GC, shown here as a lieutenant, was one of thirteen members of the Bomb Disposal service to be awarded the George Cross during the war. Colonel Archer dealt with numerous hazardous incidents and contributed greatly to countering enemy bombs and fuzes, putting his life at risk on many occasions.

Above: The George Cross was instituted in 1940 for civilians and members of the armed forces, for actions for which purely military award, could not be granted. It ranks second only to the Victoria Cross. It consists of a plain boarded cross in silver with, in the centre, a circular disc on which is displayed St George and the Dragon surrounded by the words 'FOR GALLANTRY'. In the corner of each limb of the cross is the Royal cipher. Thirteen members of bomb disposal units won George Crosses during the 1939–45 War. That illustrated was awarded to Lieutenant Ellis Edward Talbot who dealt with one of the earlier bombs that contained an unknown type of fuze.

Above right: The George Medal was instituted at the same time as the George Cross in 1940. It is awarded for acts of great bravery, and was intended as primarily a civilian award but can be awarded to service personnel in circumstances when a purely military award would not be granted. Bars are awarded for further acts of bravery. The reverse depicts a mounted St George slaying a dragon, and the legend 'THE GEORGE MEDAL'. That in the set illustrated was awarded to Major Arthur Hartley for the disposal of a wartime bomb found in particularly hazardous circumstances in 1959.

28 | UNORTHODOXY

Sappers with a taste for the more destructive side of their trade were naturally in demand for the less orthodox aspects of warfare. These burgeoned, particularly in the early days of the war, graduating from the early 'cloak-and-dagger' escapades of suspect value to taking their place as part of major operations or even whole campaigns.

A successful trail-blazer was the destruction of oil installations in North-West Europe in the face of the swift German occupation, by sapper parties from Kent Fortress Royal Engineers, during May and June 1940. Some 600 million gallons of oil went up in smoke in Amsterdam, Rotterdam, Antwerp, Dunkirk, Calais, Boulogne, along the Seine valley and at St-Malo and Brest. The raiding parties were transported by the Royal Navy on a 'one-way ticket'. Many were obliged to find their own way back in local boats, the Navy being available only where the operational situation allowed.

AIRBORNE

Even as these operations came to an end, in June 1940, two days after the Prime Minister (Churchill) had authorised the formation of a corps of at least 5,000 paratroops, Major John Rock RE,* was appointed to take charge of the organisation of airborne troops and to work out a modus operandi with the Royal Air Force. By October 1940 the first gliders had been delivered and by the end of the war two airborne divisions had been created, each with two engineer parachute squadrons, and a glider-borne field company and field park.

The first British airborne raid, on the Apulian aqueduct in southern Italy in February 1941, had achieved little. However, a smart, carefully rehearsed operation a year later to expropriate the vital parts of a German radar at Bruneval (near Le Havre), succeeded perfectly, the parachute

*Lieutenant Colonel John Rock was killed in a glider accident in 1943 after having taken over command of the Glider Pilot Regiment.

Right: Lance Corporal Vic Huggett and Sapper Wally Page of Kent Fortress RE, photographed against the background of burning oil installations at Amsterdam in May 1940. They managed to 'liberate' a camera from a passing Dutchman for the purpose. The unit, under the command of Lieutenant Colonel (later Brigadier) C. C. H. Brazier, launched numerous such raids, culminating in the major one he led himself to Normandy and Brittany. After significant success, all parties eventually returned safely to Britain though with many adventures on the way.

Parachute training at Chesterfield, typical of that organised by Major John Rock RE (see text). (IWM H23005)

Right: The German radar station at Bruneval was the target of a highly successful raid by a party of six officers and thirteen soldiers from the newly formed 1st Airborne Division, the aim of which was to remove key parts of the installation to discover German progress in radar. The sapper element under Lieutenant A. C. D. Vernon had to dismantle and eventually blow up the equipment and lay mines on the approaches. All aims were successfully accomplished and the party taken off by landing-craft within two hours of landing. (IWM D12870)

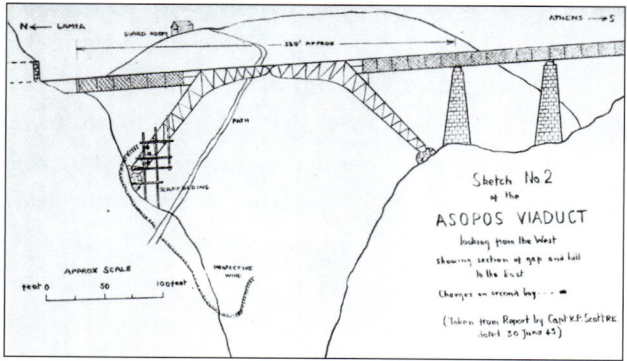

Below right: A sketch of the **Asopos viaduct** taken from the report by Captain K. P. Scott, RE.

Below: The medals of Brigadier E. C. W. Myers, CBE, DSO, who commanded the Special Operations Executive activities in Greece in 1942–3.

troops, including the sapper detachment, being taken off by the Royal Navy. By the end of 1942 an even more ambitious plan was under way to undertake sabotage in Greece by a party under Brigadier Edmund (Eddie) Myers, dropped in by air and later supported by air landing. This led to the dramatic demolition of two bridges on the main railway between Salonika and Athens at Gorgopotamos and Asopos, and the subsequent mounting of widespread harassment of the German occupation force.

The first formal overseas parachute operation was in North-West Africa (Operation *Torch*) when, using Dakotas for the first time, 1st Parachute Squadron undertook two commitments with 1st Parachute Brigade.

Against these successes has to be seen the disastrous attempt in November 1942 to set back

Hitler's atomic warfare potential by a glider-borne raid on the heavy water plant at Vermork in the Telemark district of southern Norway. After careful training in great secrecy, both of the gliders crashed and forty-one men lost their lives either in the crashes or later by being put to death by the Germans in one of the war's worst atrocities. Post-war retribution was taken.

The Chindits' operations in Burma in early 1943 and 1944 were also made possible by air support (see also page 159). There were no sapper units in the first but one of the columns was led by Lieutenant Colonel Mike Calvert with such success that he was given command of 77th Brigade for the second operation in which another of the brigades (23rd) was commanded by Brigadier Lance Perowne, both being sapper officers. Each brigade had a field company in support that performed valuable service helping the columns move through the jungle and keeping the landing grounds in the bases active. One (54th) even had to develop and operate a seaplane base for the evacuation of casualties.

COMBINED OPERATIONS

The Directorate of Combined Operations was formed in 1940 to coordinate such operations and the special training and equipment needed. The Bruneval raid (above) was one of their successes. Another was the attack on the 'Forme Écluse' lock at St-Nazaire in March 1942 in which HMS *Campbeltown* rammed and demolished the lock gates by means of a mighty delayed action explosion devised by Captains Montgomery and Pritchard.[*] The Combined Operations Pilotage Parties were another

---

[*]Sergeant Tom Durrant RE was awarded a posthumous VC for his outstanding gallantry during the withdrawal from St-Nazaire when the vessel in which he was embarked was attacked by a German light destroyer. He was mortally wounded while defiantly manning a machine-gun.

Right: One of the heavy water cells from the plant at Vermork that had been the target of Operation *Freshman* (19 November 1942) (see text). The factory was eventually put out of action by Norwegian partisans.

Centre right: Chindits crossing one of the smaller rivers into enemy-occupied Burma. (IWM IND2290)

Bottom: *ML 217.* In 1942 vessels of this type (Fairmile 'B' Class) carried the main striking force on the St-Nazaire raid. Sergeant Tom Durrant was aboard her sister ship ML 306 when it was attacked by a German light destroyer. In the ensuing battle he was mortally wounded in the action for which he was awarded a posthumous VC. (IWM HU1287A)

Right: Combined Operation Pilotage Parties (COPP). A portion of the report from the January 1944 operation to the Normandy beaches. Major Logan Scott-Bowden and Sergeant Ogden Smith were the sappers who carried out this reconnaissance.

offspring of the Directorate. Designed for close inshore reconnaissance these teams were both UK-based, but also deployed in the Middle and Far East theatres. They undertook the essential investigation of potential landing beaches, performing with extraordinary courage. Teams of eleven comprised three naval officers, five ratings and a sapper officer, senior NCO and draughtsman. Reconnaissance was normally by X-craft midget submarine.

The Dieppe 'raid' (19 August 1942), in reality a two-brigade major assault on the occupied coast of France, despite its disastrous outcome, provided the opportunity to test the principles of a major combined operation. The sapper element was drawn from 2nd Canadian Division. Extensive demolitions had been planned as part of the objectives of the enterprise. These, as well as surmounting the artificial and natural obstacles on the beach, were the main sapper tasks. Of the 333 RCE who took part only 148 returned, of whom thirty-one were wounded.

Thus unorthodoxy gradually matured within the framework of combined operations to contribute to all major operations of the war, most famously in the Sicily landings, in Normandy on D-Day, at Arnhem and the crossing of the Rhine.

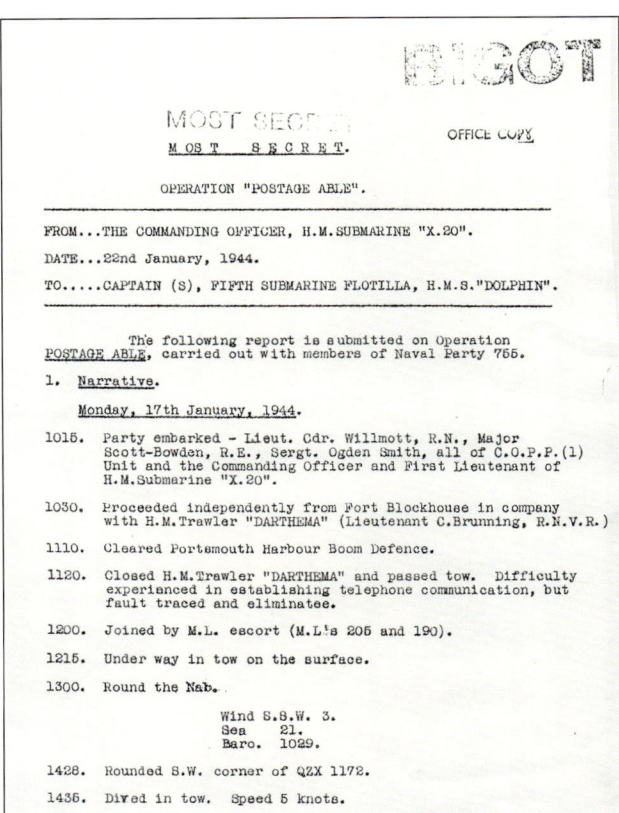

And not by eastern windows only,
  When daylight comes, comes in the light,
In front the sun climbs slow, how slowly,
  But westward, look, the land is bright.

From *Say Not the Struggle Nought Availeth,*
a poem by Arthur Hugh Clough, quoted by
Winston Churchill in April 1941, reflecting
his belief that America would enter the war

THE MIDDLE EAST AND AFRICA

The anxieties of the first two years of the war were
nowhere more apparent than in the Middle East
theatre. Italy entered the war in June 1940 and
almost at once went on the offensive in Libya
and East Africa in pursuit of her imperial expansion. The unified Middle East Command, set up
under General Sir Archibald Wavell in June 1939,
fought superb campaigns in the Western Desert
(December 1940 to February 1941) and in Eritrea
and Italian Somaliland (January to June 1941).

These victories resulted in huge losses to the Italians and briefly raised morale at home; but they
proved a false dawn as the Germans intervened
in North Africa (under Rommel) in February
1942 and in Greece in March, so forcing a dispersion of the modest British effort available.

Until this time the only British division in
Middle East Command, including East Africa and

Below: Mine warfare. Terence Cuneo's picture
Breaching the Minefields at El Alamein symbolises the
cool courage of the sappers concentrating on their
delicate task of clearing the routes for infantry and
armour in the uproar of battle. Minefields, normally each
about 200 yards deep, were laid in belts throughout the
whole five-mile depth of the German position. They were
mixed anti-personnel and anti-tank and were often fitted
with anti-lifting booby traps. Early experiences in the
Western Desert led to drills and techniques that became
part of normal life in every theatre of war. Formal
breaches as depicted were only part of the story and all
arms had to become aware of the menace of nuisance
mines that were to be encountered in every tactical
situation. Mechanical and explosive methods of clearance
were developed. The remains of one of the earliest, the
Scorpion flail, can be seen in Cuneo's painting.

Right: Eritrea was the scene of some of the toughest fighting in the campaign of 4th and 5th Indian Divisions under Lieutenant General William Platt. Indian sappers are shown here at Fort Dologorodoc, dominating high ground to the left of which lay the road block on the approaches to Keren that 5th Indian Division Engineers cleared, company by company in turn, under frequent shelling, to allow for the advance in March 1941.

the Western Desert, was 7th Armoured Division, commanded by the redoubtable former sapper, Major General Percy Hobart (see page 163). The remainder amounted to two Indian and two Australian divisions, one from New Zealand and, two from South Africa with two more made up from the West Africa Frontier Force and KAR brigades from East Africa. Most of these divisions had three organic field companies and a field park company, but 7th Armoured Division had only one field squadron (2 (Cheshire)) and a field park troop (141). There were also field companies at Corps and Army troops level.

For fifteen months from March 1941 to June 1942, Middle East Command was forced into a hand-to-mouth existence as the reinforcements of men and weapons which arrived in the theatre were dissipated to meet threats that, although important to the main prosecution of the war, were peripheral to the main business of defeating the Axis forces in North Africa. Greece (April–May 1941), evacuated partly through Crete, absorbed three divisions and two brigades of which only 1st Armoured Brigade (with its field squadron) was British. Only about half returned. In April 1941

forces had to be sent to Iraq where the pro-fascist prime minister, Rashid Ali, was besieging the British air base.* In June another force had to advance into Lebanon and Syria to force a Vichy French force out of the area. In August a British and Indian force occupied Iran in conjunction with a Soviet Russian invasion through the Caucasus, to secure the oilfields and provide a supply route for the Red Army, now under severe pressure from the German invasion of their country that had been launched the previous June.

Meanwhile, Rommel seized the opportunity of this Allied weakness and the 'see-saw' war began across Cyrenaica that was to last a year. Four major battles took place. First was the rapid Allied withdrawal in the face of the Afrika Corps' lightning first offensive (March–May 1941). The second, Operation *Battleaxe* (15–17 June 1941), was a counter-attack that failed with heavy loss. After the third, Operation *Crusader* (November 1941), Rommel was forced to withdraw to recover after three weeks of heavy fighting against the newly christened Eighth Army, in which both sides had suffered badly. In the fourth, the following June, the Allied defensive battle (Gazala, 26 May–14 June 1942), in chaotic fighting south of Tobruk, failed to stem the tide of the renewed German offensive. Tobruk, so important to the Axis operations for their re-supply, fell to them on 21 June 1942. Thirty-two thousand Allied soldiers were taken prisoner including 19,000 British.

*The sapper Colonel (later General Sir) Ouvry Roberts, at the time GSO 110th Indian Division, was sent to Habbaniya to organise the defence. He found two ornamental cannon outside the RAF Officers' Mess that turned out to be 4.5-inch howitzers. A few spare parts were acquired and the guns were brought into action with great effect. Roberts's reward was a DSO and the command of a brigade.

SAPPER STANSFIELD'S WALK

During the battle of Alam Halfa Sapper Stansfield was attached to a picket of the Royal Sussex Regiment whose task was to close one of the minefield lanes that had been left open for patrolling. Holes had been prepared some 100 yards in front of the position and the mines stacked nearby. An enemy tank approached and killed or wounded the infantrymen who had rushed out to deal with the situation. Sapper Stansfield walked deliberately forward, being fired at by the main armament of the tank, now 200 yards away as, presumably, its machine-gun had jammed. He picked up the mines one by one, armed them and placed them in their holes across the lane. While he walked back up the lane the tank stopped firing. As the CRE reported later, this may have been due to a number of reasons but perhaps it was a tribute to a brave man. He was awarded the DCM.

Above: Airfield Maintenance in the desert, improvised means!

General Auchinleck, who had replaced Wavell after Operation *Battleaxe*, now took personal command of Eighth Army and led them back to his chosen blocking position at El Alamein. From there he launched an attack on Rommel's advancing force in the first battle of Alamein (10–22 July 1942). It checked Rommel but at great cost.

There was no further British retreat. By this time Eighth Army had undergone two changes in command, culminating in the arrival of General Bernard Montgomery, with General Sir Harold Alexander replacing Auchinleck as Commander-in-Chief. With the firm base built up in Egypt, the new spirit of command and the successful defensive battle of Alam Halfa (30 August), the stage was now set for the main drama of the battle of Alamein (23 October to 4 November 1942) and its consequent operations that led to the final surrender of Axis forces in North Africa on 12 May 1943.

For Alamein Montgomery had four armoured and the equivalent of nine infantry divisions under his command. The engineer support amounted to thirty-eight field squadrons and companies* and more than twenty specialist units at Army level, including three field survey companies. Behind the Army was a huge organisation of transportation, port operating and construction, works and stores essential to maintaining the logistic support. The Chief Engineer was Brigadier F. H. Kisch who

*Field and field park squadrons in armoured divisions.

Above: A German jumping mine being dealt with in the Western Desert.

had been in theatre since 1940, originally as Chief Engineer British Troops in Egypt. He was killed in action in April 1943.

As Eighth Army advanced from Alamein, the first Anglo-American amphibious operation took place, on a grand scale, far to the west. Operation *Torch* (8 November 1942) landed forces at Casablanca, Oran and Algiers, the latter designed to seize Tunis and cut off Rommel's Afrika Corps. The new First Army, commanded by General Sir Kenneth Anderson, contained the British element. The Germans responded quickly and the campaign, initially successful, became a tough slog in mountainous country and deteriorating weather. Nevertheless, five months after Alamein the key battles for Tunisia were under way on the Mareth Line, and the Eighth Army linked up with II (US) Corps on 7 April 1943. The Axis forces surrendered a month later.

The huge distances to be covered and the difficult terrain led to a far wider dispersion of engineer units and an even greater emphasis on logistic engineering than in the Western Desert. This was recognised in the variety of specialist units such as transportation and airfield construction companies. On the field engineering side, this campaign saw the first use of parachute troops and the introduction of the Bailey bridge. (See also pages 161 and 163.)

Left: General Sir William Dobbie, GCMG, KCB, DSO (1869–1964) was Governor and Commander-in-Chief of Malta from 1940 to 1942 throughout the period when the island was under violent attack by the Luftwaffe and cut off from outside help. He had been called from retirement to take on this task. His dauntless courage, founded in an openly expressed faith in God, inspired the people of the island to the valiant defence that won them the George Cross. In his earlier career he had served in the South African War and on the Western Front. As GSO1 Operations at GHQ he signed the order to cease fire on 11 November 1918. Between the wars he commanded an infantry brigade in Cairo whence he went to Palestine to deal with a serious outbreak of hostilities between the Jews and the Arabs. He was also Commandant at Chatham and GOC Malaya before he retired in 1939, after which he became Governor of Gibraltar.

Above: A German 88mm gun at Alamein, a pen and ink drawing by Major E. Bainbridge Copnall of 85 Camouflage Company.

Left: The Africa and Italy Stars, two of the eight similar campaign stars awarded for the different theatres of operations. The stars were of bronze and have six points. In their centre is the Royal Cypher, surmounted by a crown and surrounded by a circlet which bears their name.

Right: 'Nine Sapper' Bridge, on the Florence–Forli road, Italy 1943, a painting by Lieutenant Colonel E. D. Lyons, a fine example of the versatility of the Bailey equipment (see also page 163).

SICILY AND ITALY

The Allies now carried the war on into Sicily and Italy, but the priority for planning and resources switched to the opening of a new front in Western Europe, demanded by the Soviet Union fighting for their life at Kursk and with Leningrad still besieged; and by America, seeking an early defeat of Germany through a more direct attack than offered by the Mediterranean theatre, in order to free their forces for the Pacific war against Japan.

Nevertheless, from the air- and sea-borne invasion of Sicily (July–August 1943) followed by the landings at Salerno (9–16 September) many lessons were learned that directly fed into the planning for the eventual Western European campaign. Italy, however, presented daunting new problems for the fighting troops, and particularly for engineers. The rugged, mountainous terrain of the central spine from which flowed numerous rivers prone to flash flooding into the coastal plains, greatly favoured the defence. Soon after their successful landings in Calabria and at Salerno by Eighth and Fifth Armies respectively, the Allies became bogged down against the Gustav Line (a series of lines in great depth), the first of the strong defensive positions at which the Germans hoped to hold them. Assault bridging and dealing with the plethora of mines became the standard fare of the field companies. As the infantry worked their way into the mountains, tracks were needed for their logistic support and for the movement of artillery and armour. The key to the defence was the mountain complex topped by the monastery of Cassino, which brooded threateningly over the maze of valleys and rivers that guarded it. By a process of attrition and dogged resolve, the Allied armies gradually drew the net tighter on their enemy.

Not until May 1944 were the Allies in a position to make the final breakthrough, supported by the slightly laborious progress of the January

Anzio landings. In Operation *Diadem*, which ended with the fall of Rome on 4 June, twenty-five divisions took part of which five were British.* Two bridges over the Rapido made sapper history. 'Plymouth' bridge was the first Bailey assault bridge, 100 feet long; it was carried on two Sherman tanks and laid early on 12 May by 8th Indian Division Engineers. That night 'Amazon' bridge, immortalised in the Cuneo painting (see page 161), was built under constant fire by three field companies of the 4th Division.

In the face of the Allied pursuit, the Germans conducted an orderly withdrawal to their next line of defence, the Gothic Line, running from Pisa across to south of Rimini. Allied progress was slowed and sapper life complicated by countless demolitions and mines that barred their way and by the loss of seven divisions to prepare for the forthcoming invasion of southern France. Nevertheless, they reached their objective and actually broke through as far as Ravenna when the atrocious weather brought operations to a halt for the winter.

The final decisive battle was launched on 9 April 1945 when Eighth Army with eight divisions broke through the flooded area north of Ravenna, using the only relatively dry route, the 'Argenta Gap', and crossing the Rivers Senio and Santerno before advancing on to the Po in conjunction with Fifth Army's thrust to Modena

*The remainder were US (7), French (4), Indian (3), Canadian (2), Polish (2), South African (1), New Zealand (1).

and Verona. All winter the sappers had been fully extended improving the communications in this area. Now the nature of the obstacles, full rivers with high flood banks protected by extensive minefields, called for supreme effort. By now assault engineers had arrived and proved their worth with ARKs and bridgelayers. In a brilliant operation the Argenta Gap was outflanked and the rivers crossed with the support of massive aerial and artillery bombardment. After ten days' fighting, further resistance was impracticable (see also page 172).

The campaigns in Italy, though often overshadowed by the higher priority of Western Europe, had a marked effect on the overall conduct of the war by reducing the German ability to reinforce either of their other main fronts. The appalling conditions in which it was fought are often forgotten. The sapper record can be illustrated by some statistics:

Above left: Monte Cassino, a painting by Lieutenant Colonel E. D. Lyons. The monastery (see page 155) and its approaches had been reduced to rubble early in 1944. A special engineer force, Cassino Task Force, comprising mostly South African units, was created to clear the way through the area. Three field and three road construction companies were allocated with supporting transport and pioneers.

Above right: Italian wooden mines.

Bailey bridges built	2,832	amounting to 45 miles; longest 1,126 feet
Bailey pontoon bridges	19	longest 1,100 feet
American treadway bridges	101	longest 1,125 feet
Folding Boat Equipment		6,954 feet
Permanent bridges, steel and trestle	204	
Permanent bridges, brick and masonry	73	
Airfields constructed	184	
Airstrips constructed	249	
Railway track rehabilitated		2,450 miles
Railways bridges reconstructed	490	

30 | EBB AND FLOW IN BURMA

Above: Fighter Taking Off, a watercolour by James Fletcher-Watson, who served in India and Burma during the war.

While 1942 was bringing glimpses of daylight in North Africa, Burma still lay in darkness, with essentially two corps guarding about a thousand miles of the rugged Indian border on three fronts. These were the Arakan coastal region, based on the port of Chittagong; the central region in East Bengal and Assam centred on Imphal; and the northern region based on the Chinese Yunnan province, an American responsibility, with the main aim to maintain communications (the Ledo road and a pipeline) with China.

The remains of the exiguous force, based on 17th Indian Division, that had so gallantly fought its way back into India, were gradually restored and reinforced. That force, though reduced by casualties and disease to a mere 12,000, had achieved miracles not only in rescuing itself against dreadful

Above: The Imphal–Tiddim–Kalewa Road (Imphal–Tiddim section illustrated), across some of the most daunting terrain in the region, had been completed in 1943. Its route included the famous 'chocolate staircase', a three-mile stretch involving a 2,000-foot climb to Tiddim village through thirty hairpin bends. The road was held by 17th Division when the Japanese attack towards Imphal started in March 1944. It was severely damaged in that campaign and after the subsequent Japanese withdrawal, but became the main axis of advance of 5th Indian Division to the Chindwin in 1944.

Above left: A tense moment on the Tiddim–Kalewa section in March 1944 as sappers await the rearguard before blowing the road away at milestone 11½.

Left: A Bailey catwalk built to negotiate a damaged section of the road alongside the Chindwin River.

odds but also in delaying the enemy build-up through the damage they had inflicted on the Burmese infrastructure. Every major bridge in the country had been destroyed, some of which were never rebuilt until the British Army returned.

As to future operations, Burma presented a greater logistic and engineering test even than Italy. The only one it was possible to launch at this stage (winter 1942, about the same time as Eighth Army was racing through Cyrenaica) was a divisional-sized advance into the Arakan with the aim of establishing a base from which it might be possible to mount an assault on the strategically important Akyab Island. It failed in the face of the swift reaction of the far more experienced Japanese, and 14th Indian Division retired to a position south of Chittagong.

While the first Arakan campaign was still in progress, the trail-blazing first 'Chindit' expedition (77th Indian Infantry Brigade under Brigadier Orde Wingate) penetrated behind Japanese lines across the Irrawaddy, gaining enough success to earn their promotion to divisional size for future operations. (See also page 150.)

Fourteenth Army under Lieutenant General 'Bill' Slim was created in October 1943. During that year efforts were confined to planning and training for future operations based on the two main fronts: Arakan with XV Corps, and the Kohima and Imphal area with IV Corps. For the sappers this meant all the usual support work in the forward areas with an emphasis on creating routes forward out of the mule tracks which were all that existed formerly. A special engineer force, the General Reserve Engineer Force or GREF, was set up

Top left: Grub Bridge across the Chindwin (page 160), built by 474 Army Group Engineers (67, 76 and 361 (Bengal) Field Companies and 332 (Bengal) Field Park, all of them Indian Engineers) under Colonel F. Seymour-Williams, after whose small son it was named. Following a major logistic operation to bring the equipment forward 300 miles from the railhead, including the building of twenty-two timber and nine Bailey bridges on the final 26-mile stretch alone, the bridge was started on 6 December 1944 and the first vehicle crossed at 3.00 p.m. on 10 December.

Centre left: Laying Sommerfeld tracking.

Left: The Ledo Road. (IWM4409)

to create the logistic infrastructure in Assam and Eastern Bengal that would be needed. This mundane title conceals the significance of an organisation on which the army's future was to depend. It expanded enormously as it took on control of Transportation and Movements, acquired attachments, such as from the Indian Tea Association who controlled most of the labour in the area, and other resources such as quarries. Aggregate became a critical material as GREF created the airfields, with their hard-standings and accommodation that were to be used particularly by the USAF flying over the 'Hump' in support of China. The road-building programme was also extensive and not only in the rear areas. GREF also operated under Fourteenth Army on road projects outside the capacity of the formation engineers. They also built more than 500 miles of oil pipeline.

Despite these preparations, by early 1944 operations were possible only with limited objectives. The Chindits, now divisional size, were to be thrust into northern Burma. There they would support the southerly advance by the Chinese under their American commander, General Joe Stilwell, by causing as much havoc as possible in the Japanese rear areas while IV Corps made a limited advance across the Chindwin. In the Arakan XV Corps would move south to secure

the passes over its spine, the Mayu mountains. This latter offensive came first, in January; that from Kohima in March, preceded by the launch of the Chindits in February. In both sectors they clashed violently with almost mirror-image onslaughts by the enemy.

The resulting battles were among the most critical and hard-fought of the campaign. In both, despite being surrounded, the British, Indian and West African troops of all arms stood firm. By their aggressive action they demonstrated the high standard of training and morale with which they had been imbued in the months of preparation. The Arakan battle was won just in time for reinforcements to be flown across to relieve the crisis at Imphal and Kohima where the very security of India was being fought for in bloody hand-to-hand struggles that have become epics of history. The successful outcome of the fighting on this front was the turning-point of the campaign.

Equally epic were the experiences of the Chindits. There had been no engineer units in the first expedition although several individuals had participated. The second was a five-brigade divisional operation complete with supporting arms and services. The general plan was constructed round bases from which operations could be conducted and which could allow for re-supply and casualty evacuation. Two of the brigades had special tasks as flanking forces to Fourteenth Army and Stilwell's Chinese. The other three operated mobile columns to harass the Japanese rear areas. Eventually, as casualties took their toll, these concentrated and came under Stilwell's command in support of his operations against the key centre of Myitkina. They were finally withdrawn in August and, Wingate having been killed in an air crash, they never reformed. This huge operation, dependent on resupply by air, absorbed disproportionate logistic resources from Fourteenth Army and the whole range of engineer skills was exploited. (See page 150.)

After the victories at Kohima and Imphal new plans were put into action for the liberation of Burma. Ahead lay the mighty rivers of the Chindwin and Irrawaddy and their tributaries,*

CHINDITS
Brigadier Mike Calvert, who led in the most successful of the columns of the first Chindit expedition and commanded 77th Brigade for the second. In that operation Brigadier Lance Perowne commanded 23rd Brigade.

Brigadier J. M. Calvert, DSO

Major General L. E. C. M. Perowne, CB, CBE

*Themselves often major obstacles. The Manipur River, tributary of the Chindwin, was 330 feet wide with an 8-knot current.

divided by razor-backed mountains across which led minimal tracks towards the river plains. A third corps (XXXIII) had been formed at the start of the Kohima battle and this led the advance to the Chindwin in October 1944,

Right: A Parachute Smock.

Right: A Cutting – Thai–Burma Railway, a sketch by Ronald Searle, better known in post-war years as the creator of the St Trinian's schoolgirls and the 1950s' schoolboy Nigel Molesworth. He served with 287 Field Company and was a prisoner-of-war for three-and-a-half years.

before the end of the monsoon. The leading troops had to force their way through against the enemy, who disputed every inch, against the weather, which quickly reduced tracks to mud and swept away bridges and culverts in flash floods, and against equipment and material shortages.

A bridgehead having been established across the Chindwin at the end of November, using assault boats and Folding Boat Equipment, the main crossing was established in the XXXIII Corps area. At 1,100 feet, it was the longest Bailey pontoon bridge to have been built in any theatre in the war at that stage (see page 158).[*]

By the middle of January both IV and XXXIII Corps had established bridgeheads over the Irrawaddy in accordance with General Slim's masterly plan to trap and destroy the bulk of the Japanese army north of Rangoon. Several widely separated crossing sites were used by the different formations requiring the necessary equipment to travel the same long haul from the railhead, although some lighter items were brought forward by air. The fighting that resulted from this successful manoeuvre continued until 1 April. It included the capture of Mandalay and the important command and administrative centre of Meiktila. There was more fighting to come but, in the words of the official British historian, with the capture of Meiktila the Japanese army in Burma 'had virtually ceased to exist as a fighting force'.

Meanwhile in the Arakan, XV Corps had advanced and captured the important islands of Akyab and Ramree.[†] From there, assaults were launched on the mainland to prevent the Japanese from diverting troops from the Arakan to the support of their beleaguered forces in Central Burma. It was also from there that XV Corps was able to mount its final air- and sea-borne landings to attack Rangoon from the south, just as the monsoon broke, while Fourteenth Army advanced from the north. Rangoon fell on 5 May 1945.

[*]Approximate. Various different lengths have been recorded for this bridge, which was only exceeded in length in north-west Europe.
[†]Command of XV Corps in the Arakan had been transferred to HQ Allied Land Forces for the final stage of the campaign.

31 | FORGING THE TOOLS

Although little progress in modernising the Army had been made in the 1920s and 1930s, once the country was on a war footing advances in technology led to unprecedented changes across the board in the equipment that became available and the new skills that all soldiers had to learn, but particularly sappers. Mine warfare is one example, advancing from the improvisations of 1914–18 to a whole new dimension that had to be embraced by all arms, led by the Corps. Another is the way in which air support became an integral part of operations. Airfields, with all their support facilities, were needed quickly close to the often fluid scene of operations.

Among all the new technology and equipment that proliferated as the war progressed, as far as the Corps is concerned two stand out

Left: Sir Donald Bailey, Kt, OBE (1901–85), designer of the eponymous bridge. The simple concept of a through bridge based on a ten-foot panel to be carried by six men led to the most versatile military bridging equipment in history. Its contribution to the prosecution of the war is incalculable but Field Marshal Montgomery declared that he could never have maintained the speed and tempo of forward movement either in Italy or north-west Europe without it. Donald Bailey spent most of his career at EBE/MEXE at Christchurch and became Dean of the Royal Military College of Science in 1962.

Left: Amazon Bridge. On the night of 12/13 May 1944 a crossing over the Rapido was needed in the 4th Division sector during the final Allied operation to break out of the Gustav Line. Three field companies (7, 59 and 225) took part. The site was under fire throughout the operation. Work started at 5.45 p.m. and the bridge was complete by 5.30 a.m. Casualties were fifteen killed and 57 wounded. (Painting by Terence Cuneo)

as specially significant to its conduct: bridging and specialist armour. Both had their origins in 'MEXE' at Christchurch (see page 118).

BRIDGING

By the outbreak of war the Corps had managed to bring into service a workable range of bridging equipment. There was an infantry assault bridge, based on kapok floats in the style of the cork or petrol-can float bridges of 1914–18. The folding boat equipment for 5-ton loads had been introduced and proved indispensable world-wide (see illustration, the Seine crossing). Heavier loads could be catered for by versions of the small and large box girder bridges and by the Hamilton Unit Construction Bridge and various stock span bridges. All were heavy and slow to construct and the hunt was on for a practicable solution for bridging in the field. A brief flirtation was made with the old Inglis bridge, but it proved to have reached the limit of

its potential for further development. Thus was born the Bailey bridge. By the end of the war more than 490,000 tons of Bailey equipment had been manufactured representing 200 plus miles of fixed and 40 miles of floating bridge.

The extraordinary versatility of Bailey equipment made the bridge also adaptable for wet crossings on pontoons, but rafting called for lighter solutions tailor-made for quick construction, and this need was met by the Close Assault and Class 50/60 rafts.

SPECIALIST ARMOUR

None of this, however, offered a realistic answer to bridging in the assault. Although Major Giffard Martel had foreseen and studied the problem in the 1920s (see page 118), and some experimental work had continued in the 1930s, tank design had moved on rapidly and the problem became urgent. Assault bridge design followed the three principles illustrated.

Right: A Bailey bridge over the River Orne built for the break-out from the Normandy beaches.

Far right: The Seine crossing at Vernon, a Class 9 Folding Boat Equipment bridge ('David') and 694-foot Class 40 Bailey Pontoon Bridge ('Goliath'), with the destroyed original bridge in the background.

ASSAULT BRIDGES were developed according to three basic principles. Two, the folding or 'scissors' bridge and the horizontal launched bridge, required turret-less tanks as their launch vehicle. The third was the drawbridge type, which could be carried on a turreted launcher, normally an AVRE.

Left: The prototype 30-foot Scissors Bridge No. 3 carried on the Churchill AVRE. (IWM 4296)

Below left: The prototype 30-foot No. 2 Tank Bridge, the horizontal launching method for the No. 2 Tank Bridge. (IWM 3984)

Below right: The Small Box Girder Tank Bridge Mark II.

Above: Churchill AVRE, (Assault Vehicle RE) the principal engineer assault vehicle, mounting the 'Flying Dustbin' spigot mortar.

Above: Major General Sir Percy Hobart, KBE, CB, DSO, MC (1885–1957) raised, trained and commanded 79th Armoured Division which in 1943 was charged with evolving, training and controlling in operations the specialised armour (that became known as 'Hobo's Funnies), needed to break through the German prepared defences in Normandy and subsequent operations in north-west Europe. He had been commissioned into the Corps in 1904 and joined the Bengal Sappers and Miners in India. He served on the Western Front and Mesopotamia in the First World War. In 1923 he transferred to the Royal Tank Corps. His indefatigable drive and strong personality led him to the command of 7th Armoured Division in North Africa in 1940. But his outspoken style and inability to suffer fools gladly led to his bizarre dismissal and a period when he languished as a corporal in the Home Guard. Reinstated in 1942, his qualities were then fully exploited to the creation of the force that so significantly influenced operations on the Western Front.

Above: The Flail Tank, also known as the 'Scorpion' or 'Crab', specially converted tank for mine clearing. After a period in Royal Engineer responsibility they were handed over to the Royal Armoured Corps.

Bridging was by no means the only assault equipment needed. Even before the urgency of invasion planning and the lessons of Dieppe gave the issue a boost, some experiments had been made with mine clearing using flails and rollers. However, in 1943 79th Armoured Division under the command of the former sapper, Major General Percy Hobart, was allotted the task of developing the necessary equipment specifically for the invasion and training the units in its handling. The division eventually comprised three brigades: 27th Armoured Brigade, responsible for manning the DD tanks;* 30th Armoured Brigade, responsible for mine clearance largely with Sherman flail 'Crabs'; and the 1st Assault Brigade Royal Engineers (four regiments each of four squadrons), equipped with Churchill AVREs mounting a variety of devices such as the snake mine exploder, fascines and the Single Box Girder bridge. AVREs were also adapted, at very short notice, to carry the 'Bobbins' of heavy matting that were needed to overcome the areas of blue clay discovered on the Normandy beaches. Later the sappers also acquired ARKs and Landing

*'Duplex drive' tanks were amphibious Shermans, equipped with buoyancy screens. They could manage up to 4 knots depending on sea conditions.

Vehicles Tracked ('Buffaloes', or 'Terrapins' in the British version). Armoured brigades carried their own bridges, six 30-foot Valentine No. 1 scissors bridges to each, including in Italy.

All these measures proved their worth, not only on the Normandy beaches but also in the subsequent bridgehead battles and beyond (see page 165). Assault engineers also became much in demand in Italy where an assault brigade eventually grew from a combined RE and RAC regiment formed in 1943.

Above: AVREs in Italy mounting fascines.

Left: A double ARK (Armoured Ramp Carrier) crossing on the River Senio in support of 8th Indian Division during the advance to the Po in the spring of 1945 (see page 170).

OVERLORD

As early as the autumn of 1941, even before the United States had entered the war, exercises were being mounted under I Corps Headquarters to examine the problems of an opposed landing on an enemy coast.* Brigadier (later Major General Sir Drummond) Inglis was Chief Engineer and joined the *Overlord* team when planning began in earnest in March 1943 under the Chief of Staff to the Supreme Allied Commander (the latter yet to be named). Inglis then became Chief Engineer 21st Army Group for the invasion. Thus all the complex engineer aspects of the invasion were considered, and in the mind of one man and his team from the start.

Among these was the greatest logistic military engineering project of all time, the Mulberry harbours (see page 166) which, although outside 21st Army Group's responsibilities were to have a significant impact on their planning and the allocation of resources.

For the landings, the engineer concerns were principally with engineer intelligence, in which survey played a key role; breaching of obstacles,

*As it happened, at Minley Manor.

in which the specialists of 79th Division were closely allied; airborne and commando operations to secure critical objectives behind the beaches; and transportation to guarantee the logistic back-up. Five divisions went ashore in the first wave on D-Day, 6 June 1944, two American, two British and one Canadian, with six more to follow up. Three airborne divisions (two American, one British) preceded them. The sappers of 6th Airborne Division successfully completed their tasks in the brilliant operations to seize the bridges over the River Orne and Caen Canal, and to destroy bridges over the Dives. By the evening of D-Day, despite all the perils of the beach obstacles, the minefields in depth in the hinterland and the resolute enemy defence, the British Army was established on French soil across a twenty-five-mile front to a depth of five miles. It was August before the battle for Normandy was won with the entrapment of a major part of Fifth Panzer Army in the 'Falaise Pocket'. On the way the British had fought the gruelling Operation *Goodwood* in which a skilful German anti-tank defence had defied the advance of three armoured divisions but which had taken some pressure off the Americans.

Below: Leading the Invasion, David Cobb's interpretation of the Assault Engineers action at Lion-sur-Mer on D-Day.

Above: The Mulberry harbours, one each for the British and American assaults, were conceived in response to Winston Churchill's famous minute of May 1942 declaring that the 'piers for use on beaches must float up and down with the tide. Don't argue the matter. The difficulties will argue for themselves.' In the two years that followed, the various elements were built under civil contract. In the general view of the British Mulberry at Arromanches (above), the outer breakwater, formed initially from sunken ships and then built up with concrete caissons, can be seen protecting the inner harbour, which was some 1,300 acres in extent, two miles long and about a mile out to sea. Steel 'spud' pierheads were connected to shore by three piers, effectively floating bridges of 80-foot span carried on reinforced concrete pontoons. The whole equipment was towed out to Arromanches and began to arrive on D+1. It started to operate on D+4, and by the end of August 488,700 tons of stores had been landed, 40 per cent of the total over the beaches in that period. The American Mulberry was less successful, having been wrecked in the violent storm that struck the coast on D+13.

Left: A roadway unit under tow. Six spans, total length 480 feet, made up a tow. The special kite anchor units can be seen on the deck.

Below left: The vertical motion of the pierhead units on the spud legs was controlled by power from inboard engines.

Below: The LCT pier was separate from the roadway units.

The subsequent pursuit of the German Army brought the Allies to Belgium by early September. The Germans fought hard to delay this sensational advance. Demolished bridges and resolute rearguards lay in the way. Two operations in particular underline the sapper contribution.

SAPPERS IN DEMAND

The Seine was crossed in the last week of September in operations on three Corps fronts, prepared for well ahead with units specially trained and equipment pre-stocked. In XXX Corps area the crossing was at Vernon (see page 163). The lead division, 43rd, was allocated six field and two field park companies from Army and GHQ Troops to boost its own engineers. The assault crossing in storm boats operated by 583 Field Company in the evening of 25 August was badly shot up but heroic efforts took the infantry across. The site remained under machine-gun fire; nevertheless anti-tank weapons were taken across by day on Folding Boat Equipment rafts destined to become part of a bridge. Completion of that was not possible until 5.20 p.m. on the 26th after the intensity of the enemy fire had reduced somewhat. The first of two Class 40 Bailey pontoon bridges could similarly not be opened until late on the 27th, due to the heavy artillery fire, but some tanks were ferried across; the second was completed on the 29th.

Two weeks later, at Le Havre, there took place an exemplary operation to capture the town in which the German garrison was holding out behind a daunting barrier of minefields, craters, ditches and wire, covered by concrete

Top right: A fuel distribution point somewhere in France after D Day. The 'Pluto' (Isle of Wight to Cherbourg) and 'Dumbo' (Dungeness to Boulogne) installations in the UK were built by civil contract. Their terminals and distribution system across the Channel was a Royal Engineer task. Eleven hundred miles of pipeline were laid at the rate of 3½ miles per day.

Centre right: Allied Half-Track and Lorries Parked by a Windmill, a watercolour sketch by Sapper Frank Bryson, an established artist in civil life who served in 179 Special Company in Normandy as a camouflage modeller.

Right: A Class 40 Bailey Pontoon Bridge over the River Meuse at Maeseyck, built by 8 Corps Troops Engineers in January 1945.

machine-gun emplacements. Four assault squadrons were allocated to the leading brigade of 49th Division for the initial attack on the evening of 10 September. The casualties were severe but by morning three lanes were open. The AVREs of 617 Assault Squadron played a critical part with their petards, although many were put out of action by mines. 51st Division followed up later that night with two assault squadrons in support and by the following morning the way was clear.

In this way was a multitude of battles fought at company, battalion and brigade level as the armies worked their way north-east through the flood plains of Flanders, criss-crossed with hundreds of minor rivers and canals forming natural barriers behind which the Germans could fight. Montgomery saw the Ruhr as the Allies' next main objective, to be taken by a wide flanking movement designed to cross the Rhine between Wesel on the Dutch/German border at Arnhem.

MARKET GARDEN AND BEYOND

The engineering consequences of Montgomery's concept were colossal. Rivers, canals and dykes abounded, complicated by constricted routes through low-lying country of poor going. Nine thousand sappers were made available with 2,277 vehicles to cope with the River Maas (800 feet wide), the Maas–Waal Canal and the Neder Rijn and other lesser crossings. Success for the operation depended on the swift advance of XXX Corps to link up with airborne assaults on the bridges over the canals to the north of Eindhoven, the Maas at Grave, the Waal at Nijmegen and the Neder Rijn at Arnhem, the ultimate prize that could unlock the route to the Ruhr.

All succeeded except the last where 1st British Airborne Division, landed in the very area in which two German panzer divisions were retraining and too far from their objective to make a quick snatch. After ferocious fighting at the end of the bridge, in which eighty sappers of

Below: Assault Engineers in the attack on Le Havre, 10 September 1944. The 'Crab' flails led the way, followed by AVREs and bridgelayers, but such were the casualties that by morning only three of the planned eight lanes were open. Although the armoured attack was thus stalled, the infantry attack proceeded supported by the AVREs of two troops of 617 Assault Squadron, which played a critical part with their petards, although many became casualties from mines.

Above: Liberation medals were presented to mark anniversaries after the war. Examples in this set, awarded to Sapper Freeman, are (left to right after the four war medals): the French Normandy medal commemorating the fiftieth anniversary of the invasion, the Dutch Liberation medals commemorating the fiftieth and fortieth anniversary of the liberation of Holland.

Above: 'The Rhine Crossing'. This picture by David Cobb was commissioned to commemorate the war in Germany. It depicts the construction of a Class 40 Pontoon Bailey Bridge over the Rhine at Rees by sappers of XXX Corps over the period 23/24 March 1945.

Above: General Sir William Jackson, GBE, KCB, MC*, (1917–1999), shown in the full dress of Governor of Gibraltar, which post he held from 1978 to 1982; he holds the keys of the fortress. General Jackson was commissioned in 1937 and was one of the first sappers to come under enemy fire, in Norway in 1940, for which action he was awarded the MC. He served in North Africa and Italy, and was twice wounded and won a bar to his MC in Italy. He was on Fourteenth Army staff for the re-occupation of Malaya. After the war he held a series of influential staff appointments including Assistant Chief of the General Staff (Operational Requirements) and, eventually Quartermaster General. He commanded the Gurkha Engineers from 1958–60 and was Commander-in-Chief Northern Command in 1971. A prolific writer of military history, he was co-author of the official history of the war in the Mediterranean and Middle East and was Military Historian to the Cabinet Office from 1982 to 1987.

1st Parachute Squadron and 9 Field Squadron featured prominently alongside 2nd Parachute Battalion, all were overwhelmed. A brilliant rescue operation mounted from the south of the river by 43rd Division brought out 2,700 men but 3,800 went into captivity. The invasion of Germany and the crossing of the Rhine would now have to await a more deliberate operation.

Meanwhile there was much to be done to clear the enemy from the coastal ports, which had been effectively by-passed and particularly from the approaches to Antwerp which, although in Allied hands, could not be used until the strong German presence in the Scheldt estuary was crushed. Once again the assault engineers played a decisive role, this time also in amphibious guise for which all four assault squadrons of 5th Assault Regiment had been equipped with 'Buffaloes'. Such were the difficulties of this complex campaign conducted by II Canadian Corps that it took a month from early October until 6 November until the last objective fell and a further three weeks for a flotilla of 100 craft from the Royal Navy to clear the Scheldt of mines.

21st Army Group's approach to their final major obstacle, the Rhine, was now barred by the awkward terrain of the Reichswald, lying between it and the Maas through which lay the original Siegfried Line. The ground was water-logged and flooded where the Rhine had burst its banks. In the four weeks from 8 February 1945, described by both Montgomery and Eisenhower as probably the toughest fighting of the war, XXX Corps and II Canadian Corps bludgeoned their way through the mud, over tank ditches, through minefields, across floods and all against a resolute defence. Massive sapper effort was needed just to keep the army moving let alone dealing with the obstacles. Once again the assault engineers using their full inventory of devices, proved their indispensability.

Finally, on the night of 23/24 March, just two weeks after the last German soldier had been ejected from the west of the river, fourteen divisions crossed the Rhine under 21st Army Group command, supported by a massive airborne assault by XVIII (US) Airborne Corps including 6th (British) Airborne Division. By 26 March the bridgehead was ten miles deep and being

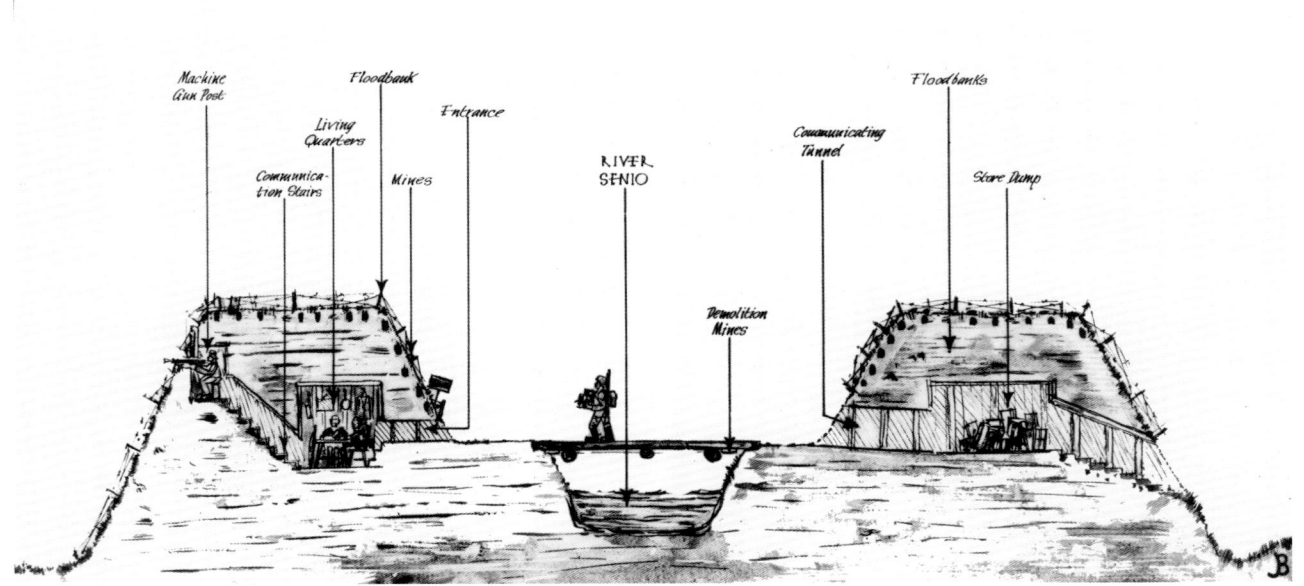

supported by numerous bridges of which six were built by British engineers, with four more completed by 5 April.

FINALE

Although there were still five weeks to go until the Germans finally admitted defeat, the Allied armies now fanned out through northern Europe in wide sweeping manoeuvres to surround and force the capitulation of the remaining pockets of resistance. In a change of policy for political reasons 21st Army Group were denied the opportunity to dash for Berlin but Field Marshal Montgomery at least had the satisfaction of accepting the formal surrender of the German forces on Lüneberg Heath on 4 May 1945.

By this time similar dramatic events were occurring in Italy. The Eighth Army Spring offensive under Lieutenant General Sir Richard McCreery was launched on 9 April 1945 (see also page 156). The exceptional complications confronting the sappers of breaking through the well-prepared German positions on the Rivers Senio and Santerno, and crossing the River Po, had been planned, trained and reorganised for over the winter. Once again the assault regiments (both Royal Engineer and Royal Armoured Corps) proved their worth both in breaching the main obstacle (see illustration) and in the subsequent follow-up where they were faced with a maze of small rivers and canals, and a proliferation of mines and booby traps.

The Germans abandoned huge quantities of vehicles, weapons and other equipment in their anxiety to escape across the Po. This was crossed by the Allies on 25 April, virtually unopposed, using storm boats and rafting. A 1,100-foot Bailey pontoon bridge was built by the evening of 27 April. A high-level bridge was completed on 4 May by South African engineers, over the remains of the piers of the former main road bridge, using parts that had been specially designed and tested over the winter.

The next major obstacle was the River Adige, reached by 8th Indian Division on 28 April. The problem now was the rate at which the speed of advance was outstripping the availability of bridging equipment. However a Folding Boat Equipment bridge, christened Roorkee Bridge, was in place by the morning of 29 April to supplement the rafts and amphibians that had taken the assaulting troops across. Two days later the Piave was crossed in similar style. The fighting was not yet over. On 30 April, a strong German force launched a determined attack on a New Zealand field park company close to the river; however, the previous day the German commanders had realised that all was lost and sent delegates to the Allied headquarters at Caserta to sign the surrender, which was effective on 2 May 1945.

In Burma (see also page 160), the beginning of April saw the Japanese forces at the end of their tether. They had been forced out of the Arakan and the strategically important Ramree

Left: The Irrawaddy Crossing by David Cobb portrays the activities of the sappers of the 2nd Infantry Division at Ngazun during the period 24 to 27 February 1945.

Left: A Class 5 Raft Folding Boat Equipment Raft crossing a river in Burma.

Island was occupied on 9 February. From there, on 1 May 1945, a sea- and airborne operation was launched by 26th Division against Rangoon in support of Fourteenth Army's southward advance down the Irrawaddy valley. Next day the monsoon broke with Slim's men still a hundred miles short of their goal but, on 4 May, 26th Division entered Rangoon unopposed.

There were now two main objectives for the Far East theatre: the destruction of the remaining Japanese forces in Burma and the preparations for the re-occupation of Malaya and

Above left: Lieutenant Claude Raymond, VC, was in command of a deception unit in Burma operating within the ill-defined enemy lines on the coast of Burma when his patrol encountered a well-dug-in Japanese detachment. He led his patrol to the attack, which he pressed home despite grievous and, as it proved, mortal wounds. His inspired leadership and self-sacrifice in insisting that other wounded members of his force received treatment first led to his award, one of only two to a Royal Engineer in the Second World War.

Singapore – Operation *Zipper*. The latter resulted in major reorganisation; Fourteenth Army was made responsible for the Malayan operations and its headquarters moved to India with a new XXXIV Corps formed for the purpose under Lieutenant General Ouvry Roberts (see page 153). In the end *Zipper* was pre-empted by the Japanese surrender, although unopposed landings were made on the Malayan coast. A new Twelfth Army was created to control all land operations in Burma. After the fall of Rangoon the Japanese forces were able to withdraw across the river Sittang and operations were conducted in the atrocious monsoon weather to prevent their concentration and push them back towards the Thai border. However, Fourteenth Army's strategy had cut off some 15,000 men in the Irrawaddy valley who were now harassed as they tried to work their way east. After surviving in the hill country between the Irrawaddy and the Sittang rivers and concentrating with the remnants of other formations, they attempted a planned breakout in July. In this final battle, lasting the whole month, the Japanese suffered some 12,000 casualties to a few hundreds from Twelfth Army.

The Japanese surrender following the Hiroshima and Nagasaki atomic bombs finally brought peace on 15 August 1945.

Right: Fighter Airfield with the Fourteenth Army, a watercolour by James Fletcher-Watson.

V
A MANY-SPLENDOURED
THING

Among the several 'families within a family' in the post-war Corps are the Gurkha Engineers offering the prospect of service among people of renowned military aptitude, with a distinctive *esprit de corps* and normally deployed on stimulating work in appealing parts of the world.

The Gurkha Training Squadron formed up in December 1948 in Kluang, Malaya, finding its manpower from re-enlisted riflemen from the pre-Independence Indian Army. Happily these

Above: Major General L. E. C. M. Perowne, CB, CBE (1902–1982), the first Colonel of the Gurkha Engineers, had earlier in his career held a series of brigade command appointments (nine commander's pennants adorned his home) (see also pages 150 and 159), and in 1952 he was appointed Major General Brigade of Gurkhas and 17th Gurkha Division during the Malayan Emergency. Nonetheless, he was a sapper through and through with experience in survey and construction and a specialist in matters mechanical and electrical. He nurtured the Gurkha Engineers in their early years with powerful argument against the many doubters of the technical potential of Gurkha soldiers.

came from all the old regiments, thus producing a good mix of the Nepalese clans and a wide spread of the old Gurkha traditions. Four difficult years followed as the fragile organisation struggled to find its identity and prove its potential. Progress was frequently knocked off course by operational necessities and organisational and administrative difficulties. Two field squadrons were raised, in 1949 and 1950, as Royal Engineers (Gurkha) with soldiers seconded from the rifle regiments of the Brigade of Gurkhas. The squadrons were moved to Hong Kong in 1950 to act in their proven role as infantrymen while carrying out their training as engineers. A fully fledged 50 Gurkha Field Engineer Regiment with two field squadrons took its place as the divisional engineer regiment of 17th Gurkha Infantry Division in November 1954 in Malaya. In 1955 the Gurkha Engineers were formed as an integral part of the Brigade of Gurkhas, but it was another three years before the Gurkha Engineers were affiliated to the Royal Engineers.

From these beginnings the Regiment went from strength to strength, winning its spurs on operations in the Malayan Emergency, the Borneo Confrontation, Cyprus, the Hong Kong illegal immigrants crisis and later in Bosnia and Iraq. They also served in the Falklands and the

Above: The Pipes and Drums of the Queen's Gurkha Engineers depicted by Jane Corsellis. From a modest beginning in 1953, they flourished throughout the second half of the 20th century, performing with distinction around the world until their disbandment in 1994.

*The Regimental Toast from those who have left is 'Jai QGE, baliyo rahanu hos' (Long live QGE, remain strong) to which the response from those remaining is 'Hami baliyo rahane chhaun' (We will remain strong).

174

Above: A short take-off and landing airstrip at Long Akah in Borneo under construction by 69 Gurkha Independent Field Squadron.

Right: The cap badge of the Queen's Gurkha Engineers, worn by officers and soldiers alike, has remained unchanged since the inception of the Regiment. On the grant of the royal title in 1977, a more elaborate badge was devised for use on cross belts, pipe banners and drum emblazons.

Above: Captain (QGO) Narbahadur Thapa, MVO, the first Queen's Gurkha Orderly Officer to be appointed from the Gurkha Engineers. He had been one of the original cadre of Gurkha riflemen who formed the nucleus of the Regiment at Kluang in 1948. Two of his sons (one of whom, Hukumraj, was to become the Gurkha Major) and a grandson followed him into the Regiment.

Gulf and they have left a permanent legacy of works around the world, particularly in Hong Kong, which became their home in 1970, and around the Pacific in numerous disaster relief deployments.

Their peak strength (nearly 1,500 all ranks) had been reached during Confrontation after the formation of a third field squadron. During the late 1960s and early 1970s they had to take their share of Army-wide cuts, losing a field squadron. Some recompense for this was the grant of 'Royal' status, made in April 1977 after a dogged campaign by General Sir William Jackson, a former commanding officer. The lost field squadron materialised again in the shape of 69 Gurkha Independent Field Squadron in 1981, the proud last vestige of the Gurkha Engineers after the final Brigade of Gurkhas reductions of the 20th century, until they were joined by 70 Gurkha Support Squadron in 2000.

Above: A Gurkha Engineer battledress blouse dating from the 1950s.

Right: A General Service Medal (1962) with clasp 'Malaya' and Long Service and Good Conduct Medal.

Royal Engineers Survey began with the 1747 survey of Scotland by Lieutenant Colonel (later Major General) David Watson. In the aftermath of the 1745 Uprising Watson had perceived the need for maps to help enforce the peace. Among the few civilians in his fifty-strong team was Watson's assistant (and nephew) William Roy who was shortly taken on to the Engineer establishment as a Practitioner.

Roy went on to great fame. After field experience in the Seven Years War he devoted much of his professional and private time to producing a military map of Great Britain. In 1784 his planning coincided with the proposal from the French to determine the relative longitudes of the Paris and Greenwich observatories. To this end Roy, now a major general, laid out the accurate base line on Hounslow Heath (today's Heathrow) from which, with verifying subsequent base lines elsewhere, the triangulation of the country began. Roy was also responsible for introducing training of both surveyors and draughtsmen. The Board of Ordnance Drawing Room in the Tower, the Royal Military Academy at Woolwich, the Staff College (originally at High Wycombe) and its junior at Sandhurst, the East India Company's college at Addiscombe and Royal Engineer Establishment at Chatham all became centres for the teaching of these essentially military skills. A Corps of Royal Military Surveyors and Draughtsmen had a brief existence from 1800 to 1817. With this impetus, the Ordnance Survey of Ireland was completed with astonishing rapidity from 1825 to 1846 under Major Thomas Colby.*

That the skills of Royal Engineer surveyors were in demand world-wide during the 19th Century is clear from elsewhere in this book (see pages 66 to 69). In Britain Ordnance Survey retained its military status, gradually developing into the national organisation that we know today.

*The need for this incidentally produced a welcome forty-eight new posts to the, then 193, establishment of the Corps, clearing the backlog of officers on half pay waiting for jobs.

Left: General Sir Edward Leach, VC, KCB, KCVO, exemplifies the front-line surveyor mapping in unfamiliar territory to support operations. He won the Victoria Cross while surveying during the Second Afghan War, courageously leading the defence of his party after they were attacked by hostile tribesmen. The painting by Peter Archer (below) depicts the incident.

Military needs outside Britain were provided through a Topographical and Statistical Depot of Military Knowledge in the War Office set up after the Crimean War which evolved into the Geographical Section General Staff whose abbreviated title 'GSGS' has survived on maps produced by military survey since April 1907.

It was the Boer War that had led to this evolution. In 1899 two officers, four NCOs and a sapper were sent out to provide mapping in areas where little existed beyond a few sketches of farms. By the end of the war in 1902 there were four survey sections and three mapping sections in the country. After the war the newly designated Geographical Section General Staff department became the authority on boundary

Right: Map of the battle of Minden, 1759. The dedication printed on the map includes these words: 'To Prince Ferdinand of Brunswick, this Plan of the Battle of Thonhausen [Minden] gained August 1, 1759 by His Britannic Majesty's Army, under the command of His Serene Highness over the French Army … is, with the greatest Respect, humbly dedicated by William Roy Captain of Engineers …'

Above: Major General David Watson (*c.*1713–1762), who initiated the survey of Scotland in 1747. Portrait by Andrea Soldi from a private collection.

surveys and played a pivotal part in the survey work of the colonies, protectorates and a number of foreign countries by invitation of the governments concerned.

This structure at least provided a framework for the needs of the army in the First World War, although few could have foreseen the extraordinary scale of the eventual commitment nor the significance of what survey could contribute to operations even beyond the phenomenal production of maps (see also page 121). Among the many new techniques embraced was aerial photography, which would eventually replace all traditional methods of map-making, and flash spotting and sound ranging of enemy guns, which led after the war to the formation of Royal Artillery Survey.

Fortunately the experience of 1914–18 was not all abandoned in the lean inter-war years. Survey training continued apace and, in 1931, 19 Field Survey Company was established as the regular element of the corps. Up until 1939 this company included a specialist training unit through which all men passed to the Survey Battalion RE, the holding unit for Ordnance Survey. In April 1940 the Survey Training Centre was set up under the Director of Military Survey, eventually located at Wynnstay Hall near Ruabon. After the war it moved to Hermitage, mutating ultimately into the Royal Military School of Survey in 1997.

Royal Engineers Survey went to war with 19 (Army) Field Survey Company and a field survey company each with I and II Corps of the BEF, thus forming survey directorates with each of the

Below right: The Kitchener Statuette is a replica, modelled by the sculptor, John Tweed, of the statue on Horse Guards Parade. Kitchener's career began in Survey first in Palestine and later as Director of the Survey of Cyprus from 1879 to 1882. (See also pages 69 and 92)

Right: 3.5-inch Ramsden Theodolite, an 18th-century instrument, originally the property of Major General William Twiss.

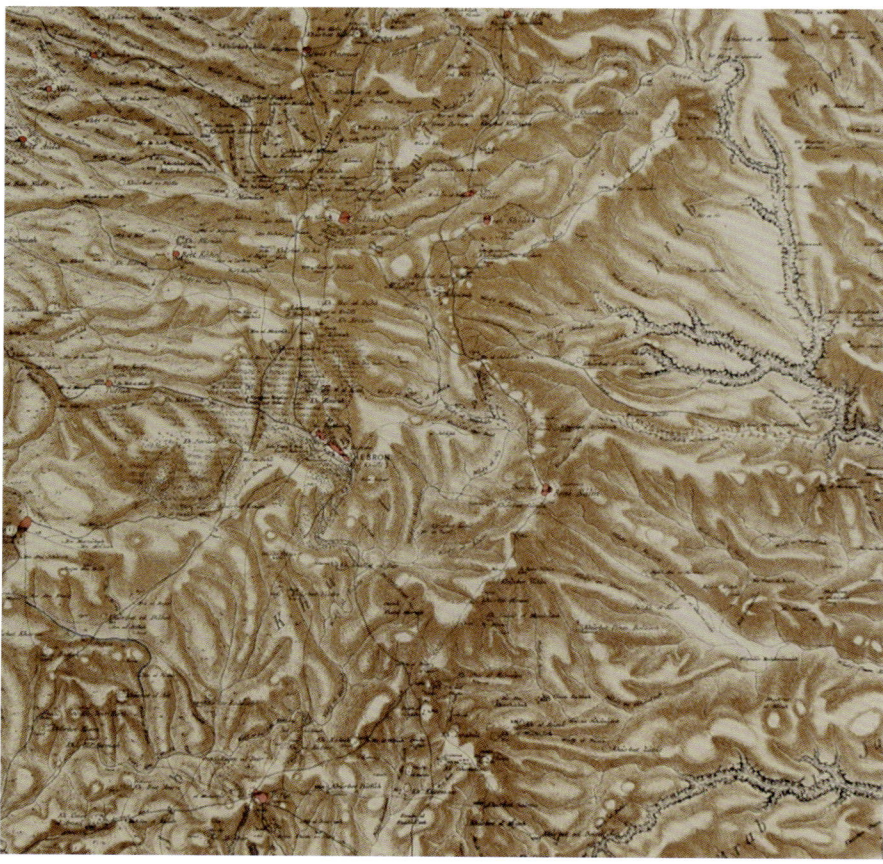

Above: A section of a map of Palestine surveyed by Lieutenant Herbert Kitchener under the auspices of the Palestine Exploration Fund, 1872–7 (see also page 69). Many of these maps were used again, with some additions, for General Allenby's campaign in 1918.

Above: Colonel Sir Charles Arden-Close, KBE, CB (1865–1952) who, as a captain during the Boer War, oversaw the first complete map production in the field – surveying, drawing and printing – by the British Army. He was Director-General Ordnance Survey during the First World War, during which nearly forty million maps were printed at Southampton for use in France and elsewhere.

headquarters. As the war expanded, a directorate, survey and printing units were established in all theatres of operation. In Burma much of the manning of these was provided from the Survey of India. The needs of the Normandy invasion were met by mapping at a scale of 1:25,000 of the northern French coastal area up to sixty miles inland, some 1,000 separate map sheets. This was the product of complete photographic cover of the area by 342 sorties by Spitfires from 140 Squadron RAF between 24 July 1942 and 17 August 1943.

The close liaison between Royal Engineers Survey and the RAF during the war continued

thereafter and assumed global proportions so that the Director of Military Survey (a major general from 1974) eventually acquired tri-service responsibilities. These were discharged through the Directorate and various specialist units in the UK. Field support to the Army was provided by field survey squadrons in UK, Germany and the Far East with 42 Survey Engineer Regiment, initially in the Middle East and returning to the UK in 1963.

Above and above left: The medals of Brigadier Martin Hotine, CMG, CBE (1898–1968). Hotine (right), who in 1941 was appointed the first Director of Military Survey, became a world-wide authority on geodetic survey. His career included a tour with Ordnance Survey during which in 1935 he organised the re-triangulation of Great Britain.

As the overseas commitments reduced, so NATO became the focus of concern in the 1970s and 1980s. Military Survey was represented throughout the command structure from the Headquarters in Mons down to brigade level. The range of products for BAOR and RAF Germany that resulted was colossal, from special road maps covering the British Forces Area to such as detailed low-flying charts. More and more the emphasis grew on the provision of mobile up-to-date geographic support to forward troops on the ground, the Tactical Information Printing System – TACIPRINT – being introduced in the late 1970s.

Outside the NATO area Military Survey had to be geared up to meet sudden emergencies, most notably the Falklands War in 1982, the Gulf War in 1991 and the Iraq War in 2003. These wars illustrate how within almost a single decade, technology had revolutionised geographic support and raised its profile. The vast increase in the

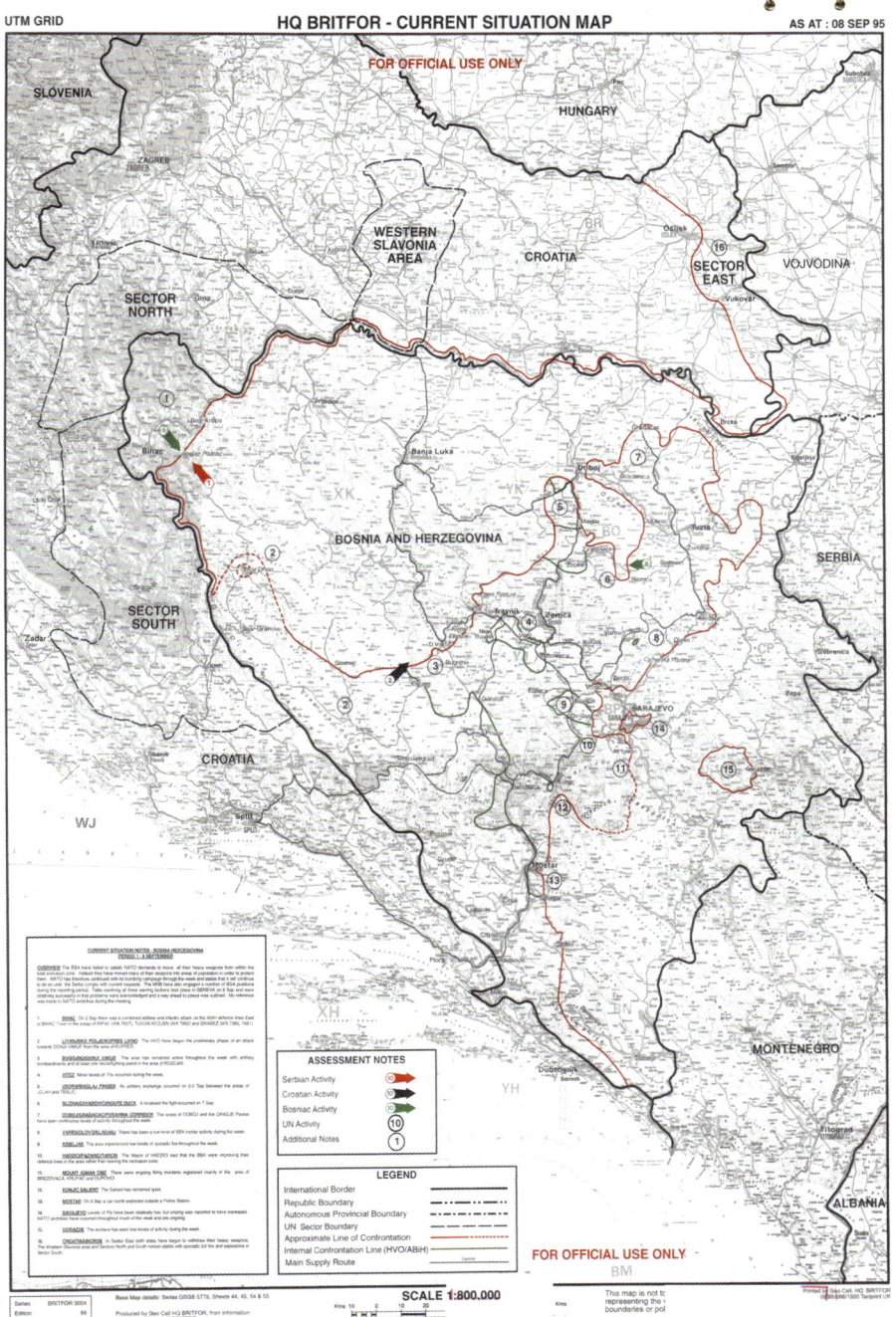

Above: A Situation map, an example of specialised mapping scale 1:800,000, produced for the NATO Peace Implementation Force (IFOR) in September 1995.

amount of geographic and terrain information available, and in the manner of its dissemination and presentation made the Geographic Engineer Group an increasingly essential part of operational capability.

Throughout its history military survey has had an inescapable international dimension, as the 1787 Greenwich–Paris link demonstrated. Its product has also become indispensable to the civil community. Hence the coalescing of the various aspects of its work, formerly seen as a military directorate, with wider government business through the Defence Geographic Imagery and Intelligence Agency to which Royal Engineer Survey is now responsible.

Above: The Military Survey 250th Anniversary Centrepiece, a replica of a Heidelburg lithographic printing press, presented to Military Survey jointly by the Corps and the Defence Surveyors' Association to mark the occasion. Printing has been a key part of Survey's operational capability since Charles Arden-Close (q.v.) set up his presses in the field in South Africa.

Right: The commemorative map of the Gulf War drawn by David Johnson, a retired cartographic technician.

OP GRANBY/DESERT SABRE 1(UK) ARMD DIV Oct 90 - Mar 91

Scale approx 1:1,000,000

LIMITED EDITION

Designed and Drawn by SSgt DN Johnson RE
School of Military Survey

35 | DANGER AND NECESSITY

The menace of unexploded enemy bombs was not faced up to in Britain until the German invasion of Poland in 1939. In November of that year, bomb disposal parties of one junior NCO and two sappers were found and trained by Royal Engineer field companies. During the next months, in the absence of the expected heavy air attacks, the three-man bomb disposal teams tended to be forgotten and they gradually disappeared into oblivion. In February 1940 the War Office accepted responsibility for all unexploded enemy bombs except those on Admiralty and Air Ministry property. In May they issued an instruction to the Commandant of the School of Military Engineering at Chatham and the commanding officers of the four RE Training Battalions, stating that it had been decided to form 109 bomb disposal sections each of one officer and fifteen men, twenty-five

sections to be formed immediately. By June the full complement of 109 had been formed and by July the number of sections had been increased to 220. At this stage it was realised that bomb disposal was a field operation which could not be conducted from the War Office and so responsibility was handed over to GHQ

Above: Bomb Disposal badge, worn on the left forearm by all ranks who qualify as bomb disposal engineers.

Left: A 1,000kg German bomb being moved by Staff Sergeant Hartshorn and his section in Sussex c.1940/1.

Left: 'Royal Engineers Bomb Disposal Section, London 1981',** commissioned by 33 Engineer Regiment (Explosive Ordnance Disposal) depicts a typical post-war incident. To achieve maximum realism the scene was set up by a hoax call-out of the emergency response crew. The equipment shown includes a Pass pneumatic trepanner, Ford compressor and Wickham steamer. (Artist: Barry Linklater)

Home Forces on 20 August 1940. A Bomb Disposal Directorate was therefore established on 29 August 1940 under the command of Major General G. B. O. Taylor. By that time some 2,000 unexploded bombs remained to be dealt with.

His command expanded to deal with this dramatic increase, construction and quarrying companies being converted initially to meet the need both in the United Kingdom and in overseas theatres. The wartime strength rose to twenty-seven companies each of ten sections. Training, technical knowledge and the development of equipment and techniques had to keep pace with the ever-increasing ingenuity of the enemy. Awe-inspiring acts of bravery were performed in the process of investigating new fuzing mechanisms and in attempts to minimise collateral damage by premature detonation (see also page 148). Inevitably many gallant operators were killed (394 from the Bomb Disposal service as a whole) in the course of these duties, which also brought recognition in the form of thirteen George Crosses and many other decorations and honours.

Even after the bombardments of the Blitz, principally high explosive, incendiary and some parachute landmines, the United Kingdom was subjected to further enemy attack aimed at undermining public morale. The Butterfly bomb raids were an example, followed later by the more significant but still ineffective V1 ('doodlebug') and V2 rocket onslaughts. Another major task arose from the extensive beach

Above: A typical post-war bomb disposal operation occurred in July 1959 when workmen excavating alongside a sewer discovered a 250kg German bomb fitted with two fuzes. The 'steaming out' process is illustrated under the supervision of Major A. B. Hartley.

Above: The Conspicuous Gallantry Medal awarded posthumously to Staff Sergeant James Prescott (left) for his gallantry in disarming bombs aboard HMS *Argonaut* on 22 May 1982 during the landings on the Falkland Islands. He was killed while he and Warrant Officer Class 2 Phillips were tackling two more aboard HMS *Antelope* on 23 May 1982. Phillips, who was awarded the Distinguished Service Cross for his part in the affair, was severely wounded (see page 248). The Conspicuous Gallantry Medal is a very rare naval decoration instituted in 1855 and never before awarded to anyone other than a member of the Royal Navy.

Above: The George Medal awarded to Captain John Stanley Bartholomew for the attempted rescue of a girl from the sea off Great Yarmouth. A minefield had to be crossed in the process. Lieutenant Evan Whitehead, who accompanied Bartholomew, was awarded an MBE.

minefields that had been laid, not always well recorded or marked, as part of Home Defence in the early days of the war. Many casualties began to occur and, while these had to be dealt with, the more general task of clearance was also begun, a continuing commitment that remained with the service for many decades.

Above: Water jetting to expose beach mines. The minefields hastily laid around the British coast left a perilous legacy. Between 1943 and 1947 151 men were killed in the process of locating and clearing these mines whose position, depth and condition had altered over the years. This work continued into the 1970s.

Above: The Royal Engineer Bomb Disposal centrepiece depicts the render-safe procedures being carried out on a Second World War German 500kg bomb. Two plates on the plinth illustrate 'steaming out' and 'beach mine clearance', the third carrying the Corps cipher.

In the immediate aftermath of the war, Royal Engineer Bomb Disposal was regarded largely as a clearing-up operation. The Headquarters moved to Horsham with an establishment of five independent troops of ninety men and a manpower ceiling of sixty British all ranks, the remainder being made up by civilians, mainly non-German former prisoners-of-war. In the United Kingdom the mine clearance and recovery of wartime bombs continued apace. There were overseas commitments, too, particularly a major dump of Japanese ordnance in tunnels in Penang, which were only finally cleared by a special team in 1967.

The possibility of future expansion in time of war was covered, up to 1967, by three Army Emergency Reserve regiments and two Territorial Army squadrons specialising in bomb disposal. However, these disappeared in the Reserve Army review of 1966 to be replaced by a single TA specialist team based in Rochester. About the same time, the tri-service nature of the work and the need to delineate responsibilities between the Services and within the Army as between the Corps and the Royal Army Ordnance Corps, led to some sensible rationalisation. This found its form in the Joint Services Bomb Disposal School, developed from the existing RE Bomb Disposal School in 1959, and the adoption ten years later of the term Explosive Ordnance Disposal to describe more accurately the wider scope of the work that for many years had embraced ammunition clearance. The school further developed into the Defence Explosive Ordnance Disposal School and became the centre of doctrine and training for all three Services.

By the 1970s intelligence assessments of the bombing capability of potential aggressors brought home the seriousness of the situation that could occur in limited or general war against the Warsaw Pact. An operational Regimental Headquarters (33 Engineer Regiment) was established with one regular EOD and two volunteer squadrons. A further regular squadron was formed for the sole purpose of clearing Maplin Sands, the proposed site for the third London airport, which never came to fruition.

The Falklands War then prompted yet more rethinking. To tackle both the unexploded ordnance and the minefield marking and clearance after the war, 33 Regiment had to form teams from its limited pool of trained officers and senior ranks, and a second regular EOD squadron was formed. Even this proved inadequate to meet the commitments that arose in the Balkans, and by the time of the Gulf and Iraq wars there existed a regular regiment of three field squadrons (EOD) as they were now designated and a support squadron, with its three-squadron volunteer counterpart 101 (London) Regiment (EOD)(V), formed in 1988.

36 | MOVERS AND SHAKERS

UP TO 1918

At the height of the First World War Royal Engineers Transportation accounted for more than a quarter of the total Corps strength of 310,000. Happily an embryonic structure existed on which this phenomenal growth had been able to develop. The Franco-Prussian and American Civil Wars, the first major conflicts of the railway era, had demonstrated the need for a trained military staff who would be able to communicate with the civilian professionals. Kitchener's railway in the Sudan (see page 86) shortly followed by the Boer War (see page 90) had seen the earliest large-scale involvement in the business by Royal Engineers, whose experience was to prove invaluable. The whole question of control of the railways in war and the war role of civilian railway engineers and managers had been exhaustively examined under the auspices of the Engineer and Railway Staff Corps (see page 189) in the last two decades of the 19th century. In 1897 an Army Railway Council had been established.

Shortly after the Boer War the Directorate of Movements had been set up under the sapper Brigadier General Henry Lawson. Under his leadership the Transportation Centre at Longmoor was established, based on the three existing railway construction companies 8 (formed 1882),

10 (formed 1885) and 53 (formed 1899), which was to train generations of sappers of all ranks until the formation of Royal Corps of Transport in 1965. At the same time selected officers were employed in the management of the railways in a tradition that had its origins in long-established practice in India and Africa.

Above: General Lord Robertson of Oakridge, GCB, GBE, KCMG, KCVO, DSO, MC (1896–1974), will always be connected with his appointment in 1953 as Chairman of the British Transport Commission, which precipitated his retirement from the Army and prevented the possibility of following in his father's footsteps as Chief of the Imperial General Staff. In fact, he never served as a Royal Engineer Transportation officer, but the principles of this specialization would have contributed to the remarkable grasp of administration that he displayed in a series of top appointments. His distinguished First World War service earned him the DSO and MC, but between the wars he retired from the Army and worked in South Africa as Managing Director of Dunlop. After proving his talent for organization as Brigadier AQ Eighth Army, he rose rapidly to lieutenant general as Chief Administrative Officer at Allied Forces Headquarters. Later he was Deputy Military Governor of the post-war British Zone of Germany, Commander-in-Chief British Forces in Germany and was seconded from the Army as the first British High Commissioner for Germany. His final Army appointment was as Commander-in-Chief Middle East in 1950.

Right: The Transportation and Movement Control Centrepiece, presented to the Royal Engineers Headquarters Officers' Mess in July 1965 by Transportation and Movement Control officers to commemorate the transfer of their service to the Royal Corps of Transport. A dolphin and mermaid are represented, rising from the sea, on a plinth with three statuettes of a locomotive fireman, a stevedore and a lighterman. An encircling train symbolises Movement Control.

Right: 'Lord Kitchener's coach', for years a showpiece on the Shoeburyness Military Tramway and now happily acquired by the Royal Engineers Museum. Almost certainly built for service overseas in about 1880, its use by the Sirdar of Egypt remains an agreeable legend awaiting further research.

Below: A model of Z Craft No. Z50, originally presented to Brigadier (later Major General Sir) Eustace Tickell who, as Director of Works Middle East Command, oversaw the design and construction of these very successful vessels. Their purpose was to convey men, vehicles and stores on inland waters but to be able to make open sea voyages unladen. Between 1941 and 1943 eighty-eight were fabricated in India and assembled on Lake Timsah in the Canal Zone. Thirty of a modified version, called 'ZZ', were made for the Royal Navy for minesweeping.

1914–1918

By contrast there was no similar organisation in place for Inland Water Transport before 1914, when the potential of the French canal system needed to be exploited and the management of the docks on both sides of the Channel to be taken under control. A Royal Engineer Docks organisation was created as part of Inland Water Transport and both of the principal arms of Transportation came under the Directorate of Movements. As part of its responsibilities acquired from the Admiralty, the Corps developed Richborough port as a depot for cross-Channel barges and built the first electric arc-welded barge, a 225-ton vessel, for the purpose.

The problems of moving tens of thousands of men to and fro across the Channel up to their units, of logistic re-supply and casualty evac-

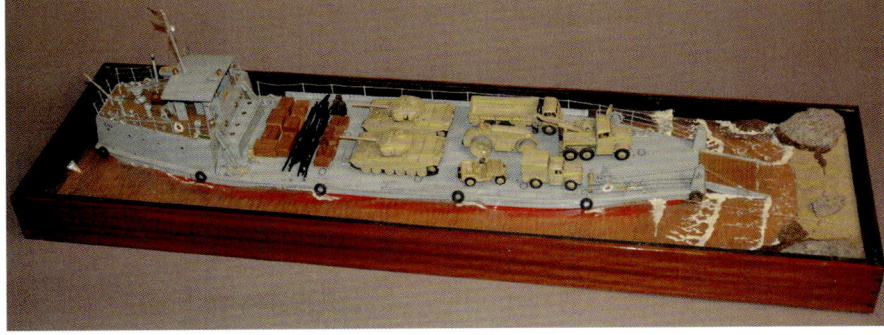

uation* in a regular programme were severe enough. However, frequently these routine movements had to be superseded by urgent operational redeployments, such as during the great German offensives of 1918.

By that time, however, the whole transportation system on the Western Front had been reorganised under Sir Eric Geddes. Geddes, who in 1914 was Deputy General Manager of North Eastern Railways, had been appointed Director General of Military Railways and Director General of Transport in the BEF in the rank of major general. In 1914 he had raised a unit of railwaymen, 17th (Service) Battalion, Northumberland Fusiliers (NER pioneers) as an entire company-sponsored battalion.[†]

By November 1918 no less than eighteen different types of transportation unit existed in the United Kingdom, such as the railway construction and operating depots at Longmoor and the Inland Water Transport units (shipyard, [port] construction, marine, train ferry, stores companies) at Richborough, Southampton and elsewhere. On the Western Front the tally was even greater with twenty-three types (including railway construction, broad gauge operating, signal and interlocking, wagon erecting and various light railway operating and technical companies).

Beyond the Western Front, transportation was controlled within each theatre of operations. The problems were more related to the distances that had to be covered, the lack of existing infrastructure and shortage of equipment rather than of coordination. The proliferation of types of company could not match the Western Front: some ten altogether, mostly railway construction and operating, in East Africa and Mesopotamia, with a multitude of small depots and detachments under their Inland Water Transport headquarters dealing with dockyard engineering, shipbuilding, local vessels and fuel supplies.

Perhaps the most impressive achievement in the transportation field in the Middle East was the railway and pipeline across the Sinai desert from the Suez Canal to Rafah in Palestine, built in stages

*The first ambulance train began to run in France in October 1914. Another twenty-nine had been provided by the end of the war.
†Geddes was recalled to England to become First Lord of the Admiralty in June 1917. He was replaced by the sapper, Major General Sir Sydney Crookshank.

as the Turks were pushed eastwards in 1916 and 1917. In the climax of this work twenty-nine miles of standard-gauge railway were built in the first two months of 1917, bringing the line beyond Rafah and providing for the logistic build-up from which General Allenby's brilliant campaign was to be launched in the autumn of that year.

BETWEEN THE WARS

The war had set transportation firmly in its place as part of the Army's essential support. The Transportation Directorate and Movements Branch of the staff was a settled feature in the Army's organisation. Transportation was now a specialisation within the Royal Engineers in which individuals could obtain experience both in military establishments and by attachment to civil firms. However, on the outbreak of the Second World War, all had to be arranged through a single staff branch of the Directorate of Movements and Quartering, QMG 13, under a colonel.

1939–45

The urgent need was the support of the BEF. On mobilisation, two dock groups went immediately to France to look after the ports of Le Havre, Brest, St-Nazaire and Nantes. Three railway construction companies and a railway survey and railway construction and operating company were all that was available at first, to cope with the necessary expansion of the French facilities. In all 141 miles of track were laid before the withdrawal.

As the war progressed every theatre built up its own Transportation organisation to repair and develop the docks,* build and manage the railways and provide movement control. In the United Kingdom two military ports were constructed at Faslane and Cairn Ryan to replace the capacity lost from the Channel ports' vulnerability to enemy action.

Apart from the Mulberry harbours (see page 166), perhaps the greatest of Transportation achievements was the support of the operations in Persia and Iraq Command ('Paiforce') designed initially for the defence of the Soviet Union border area and later committed to the

Above: A Stanier 8 locomotive, the *Corporal Lendrim VC*, victim of violence that broke out at the time of the abrogation of the Anglo-Egyptian Treaty in 1961 (see page 211). The Staniers gave great service in the Middle East where nine of them were running, all named after sapper VCs.

colossal 'Aid to Russia' programme. The development of exiguous existing facilities in a severely underdeveloped country to the point where 2,000 tons per day could be delivered by means of an 800-mile railway that ran from ports in the Gulf across desert and over mountain passes up to 10,000 feet above sea level was a truly extraordinary achievement.

POST-WAR

Britain's world-wide commitments in the withdrawal from Empire meant the retention of significant Transportation staff and units. For example, in the Middle East in 1951 the 'Tn' regiment commanded ten Railway, 169 Railway Workshops, 1,207 Inland Water Transport and 53 Port Squadrons. There were nine Stanier 8F 2-8-0 locomotives in theatre.

In 1965, after a comprehensive review of the Army's logistics, RE Transportation and Movement Control was handed over to the newly created Royal Corps of Transport. With it went the Transportation Centre at Longmoor with 8 Railway Squadron, 17 Port Regiment at Marchwood (comprising three port squadrons), another port squadron in Singapore and a lighterage troop in Cyprus. In addition, 1 Railway Group, in charge of all War Department railways, with 100 locomotives, 2,000 wagons and 645 miles of track, was handed over.

Above: Iran was re-supplied through the same considerable transportation system as Paiforce (see text). The Iranians had caused anxiety by their unwillingness to expel German nationals and in August 1941 a force, based on 8th Indian Division, was sent to secure the oilfields and eventually to occupy Teheran.

Right: Marchwood Military Port on the western side of Southampton Water.

*In the Mediterranean dock repair and development was the responsibility of the Chief Engineer's works organisation. Elsewhere it came under Transportation.

Above: The Royal Army Service Corps centrepiece was presented to the Royal Engineer Headquarter Mess by the officers of the Royal Army Service Corps on the formation of the Royal Corps of Transport in July 1965. The piece is a model of a 10-ton bridging-lorry and trailer carrying Heavy Floating Bridge equipment and serves as a tribute to the close association between the two corps and to record the formation of the Royal Corps of Transport.

Above: Tunic of the Cheshire Railway Battalion dated about 1910. This Territorial unit recruited from the London and North Western Railway. Note the collar dogs.

Left: The medals of Sapper Drawbridge, a 'Tn' soldier who served in the Middle East and Italy.

Above: A Working Model of a railway wagon used as living accommodation by Brigadier General W. V. Scudamore during the Boer War when he was operating a section of the Cape Railway.

AUXILIARY FORCES BEGINNINGS

Two 20th-century world wars bear witness to the indispensability of the Volunteer side of the Corps. Even at the beginning of the 21st century, their skills continue to prove of inestimable value, plugging gaps that exist in the diminishing Regular establishment. The story of the Army's reserves dates back to the days when the Spanish Armada threatened invasion and levies of infantry were raised. The Militia, a form of conscripted citizen army for home defence, appeared at the end of the 18th century and for some decades ran in parallel with a Volunteer force, both providing essential back-up during the Napoleonic wars. All fell into neglect after 1815. Pressure was exerted on the government by the Duke of Wellington, Sir John Burgoyne and other influential soldiers to rectify the lack of any organised form of home defence. But no serious action was taken until after the Crimean War when Parliament approved the formation of a Volunteer Rifle Corps, the volunteers to provide their own arms and cover all their expenses except while on active service.

The proposal met with huge popular response. As well as the infantry, the capabilities that particularly benefited were telegraph, submarine mining and transport. For example, Volunteer telegraphists from 24th Middlesex Rifle Volunteers supported the regular telegraph companies in the Suakin expedition of 1885. Engineer battalions formed in many counties, the number rising to fifteen by 1886. The submarine mining service, which had eight regular and four militia companies, acquired four companies appropriately trained from the 1st Gloucestershire, 1st Lanarkshire, 1st Lancashire and Newcastle-on-Tyne battalions to look after the defence of the Severn, Clyde, Mersey and Tyne respectively. Without these reservists the submarine mining service would not have been able to function. Two other militia regiments joined the Corps order of battle at this time: the Royal Anglesey and the Royal Monmouthshire Royal Engineers, the latter title surviving in the 21st-century Territorial Army (see box). Both militia units sent companies, together with fifteen sections of Volunteers, to the Boer War.

The Volunteer movement may also be credited with regularising many of the problems of transport, the lack of which had hitherto so hamstrung commanders of expeditionary forces. The Land Transport Corps, set up for the Crimean War, had served its purpose and

Below: One of a set of six candelabra presented to the Royal Engineers Headquarter Mess by members of the Territorial Army. Each has ten branches and the base bears the Corps monogram and also that of the Territorial Army.

Left: The Territorial Decoration and Territorial Efficiency Medal in this group awarded to Major R. Whitman signify long service both as a soldier and then after commissioning, a rare distinction.

Above: The Militia Long Service and Good Conduct Medal was instituted in 1904 and was superseded by the Efficiency Medal with clasp MILITIA in 1930.

then been disbanded, its place taken in the Army's Transport Branch by the 'Military Train'. At the same time the Engineer and Railway Staff Corps had been founded (see box) to bring some order to the use of railways. The Cheshire Railway Volunteers were created within the county's rifle regiment, becoming known as the Crewe Volunteers. This evolved into the 2nd Cheshire Railway Battalion and in 1885 it transferred to the Corps. Volunteers from many of the counties joined the regular railway companies both for the 1884 Nile (Gordon relief) and 1885 Suakin expeditions. They also served with distinction in the Boer War, winning a DSO and a DCM.

Above: Field Marshal Lord Nicholson of Roundhay, GCB (1845–1918), who was commissioned in 1865 and went to India. After some years with the Public Works Department he served with distinction in the Second Afghan War and was promoted Brevet Major. He went to Egypt as a field engineer and was present at the battle of Tel-el-Kebir. Further active service included the 1886 Burma campaign and the 1897 Afridi campaign as Chief of Staff to the Punjab Army for which he was made KCB. By this time he had made his mark with Lord Roberts (as his Military Secretary in 1890) who took him to South Africa on his staff and later to London as Director General of Mobilisation and Intelligence. After a spell as Quartermaster General he was appointed Chief of the General Staff working closely with Mr (later Lord) Haldane on his reforms (see text). He was promoted Field Marshal in 1911 and raised to the peerage on his retirement the following year.

THE TERRITORIAL FORCE

The Boer War, however, exposed the deficiencies of the Volunteers while at the same time pointing to the necessity of properly organised reserves for home defence, based on a county framework. The Haldane reforms abolished the Volunteers and Militia, replacing them with a Territorial Force for home defence and a Special Reserve as a 'second line' to the regular Army (see page 120). Fourteen infantry divisions and fourteen mounted brigades were established under operational command of the existing military districts but to be administered by county associations. In addition, mounted support units, Army troops and special troops for coast defence, manned

THE ENGINEER AND LOGISTIC STAFF CORPS RE (V)

is a unique unit within the Territorial Army whose purpose is to provide professional advice on engineering and logistic (particularly transport) matters. It comprises some sixty distinguished members of their professions who serve with rank but no pay, but are available to the Ministry of Defence for advice on a wide range of matters. Its origins lie in the 1860s by which time railways were playing a significant role in military operations. It was perceived by the then Honorary Secretary of the Institution of Civil Engineers, Lieutenant Colonel Charles Manby, that in the event of invasion Britain would need a body to coordinate the movement of troops and the construction of defences and that such a body could be made ready by the formation of a Volunteer Corps.

The Engineer and Railway Staff Corps, as it became, then worked closely with the Royal Engineers and the railway companies, whose senior engineers and managers formed its membership, to evolve the necessary coordination that would be needed in the event of war. Their work paid off during the general mobilisation in 1914 although by then the Army's General Staff had absorbed some of their functions. For this reason there followed a long period of stagnation until 1945 when it was appreciated what a valuable asset such a body could be to the Royal Engineers in its post-war role. From then on it thrived and developed its current inestimable role. In 1984 it became the Engineer and Transport Staff Corps RE (V), recognising the link with the Royal Corps of Transport that had been established since their formation in 1965, and in 1996, after the formation of the Royal Logistic Corps, adopted its present title.

In these different guises it has developed into a highly proactive body, an indispensable link with industry for both the Royal Engineers and Royal Logistic Corps, giving advice through its specialist liaison groups across a wide range of subjects and earning its spurs through the provision of experts and advice in recent years in the Falklands, the Balkans, the Gulf and Iraq.

from outside the divisions, were established. The responsibility for submarine mining and torpedoes was handed over to the Royal Navy.

For the Corps Volunteers all this amounted to twenty-eight field and fourteen telegraph companies for the divisions with twenty-three works and nineteen electric light companies for the coast defence work. The Royal Engineer militia units were reorganised to supplement regular units with two siege and three railway companies. On the outbreak of the First World War these Territorials were first sent abroad* to relieve regular units for service in France. However, the need to provide a third field company for the BEF divisions soon resulted in these being provided from territorial units, the first eight beginning to arrive in France in December 1914. By the end of the war about a quarter of all sappers engaged in all theatres were from territorial units.

BETWEEN THE WARS

In the 1920s the most significant event was the formation of a new category, the Supplementary Reserve, which came into being solely because of the need to provide a railway transportation force that could be mobilised more easily than the Territorial Army (as the Territorial Force was now known). The heart of it was nine railway companies (operating, construction and stores) and a docks group and operating company. They bore in their title the name of the company from which they were recruited, e.g., 150 (LNER) Railway Construction Company. The Supplementary Reserve became enormously popular, providing a field of activity for many veterans of the 1914–18 war and, as it happened, a framework for expansion in 1939. The Supplementary Reserve provided almost all the Army troops workshops and stores units at the start of the war.

As the war loomed, the Territorial Army was doubled in size, from fourteen to twenty-eight divisions. This dramatic increase threw an enormous burden on the regular Army for training and staff support. Nevertheless, on the outbreak of war the Corps element was over 76,000 in strength compared with the regular strength of 13,000.

*Notably to Egypt and India.

THE ROYAL MONMOUTHSHIRE ROYAL ENGINEERS (MILITIA)

This regiment traces its history back to at least 1539. It incorporated the first 'Royal' in its title in 1804 when it was the Royal Monmouth and Brecon Militia. By then it had been embodied four times, in 1759, in 1778, in 1793 and in 1803. It was embodied again in 1854 at the time of the Crimean War.

In 1877 it became part of the Corps on its conversion to engineer militia. During the Boer War, while not embodied, it none-the-less sent three companies to South Africa. In 1908, on the formation of the Territorial Force and the consequent demise of the Militia, it became a unit of the Special Reserve. During the First World War it provided eight companies (a total of 76 officers and 2,113 rank and file), serving variously on the Western Front, in Italy and in the Middle East.

It was mobilised again at the start of the Second World War, sending two companies (100 and 101) to France in 1939. While the bulk of 100 Company was captured at Dunkirk, sufficient escaped to enable the Company to be rebuilt in Britain. Both companies took part in the campaign in North-West Europe. It sent a squadron to Iraq in 2003.

It is the senior regiment in the Reserve Army.

Left: 107 Field Company bridging the Cidemli Dere in the First World War Salonika campaign.

Below: The Sussex Volunteers being inspected by Mr Winston Churchill during the Second World War.

Right: Posters such as this attracted enormous support on the formation of the Territorial Army.

Above: The Special Reserve Long Service and Good Conduct Medal was instituted in 1908, and awarded to NCOs and men who completed fifteen years service and attended fifteen camps. At the time of first publication of this volume, only seventeen of these medals had been awarded to members of the Corps.

POST-1945

The new Territorial Army that formed in 1947 comprised two armoured, one airborne and six infantry divisions. Thirty-four engineer regiments formed part of the original order of battle, those outside the divisions being organised in eight engineer groups. The Supplementary Reserve, later renamed the Army Emergency Reserve, which unlike the Territorial Army had no permanent base and a commitment of only two weeks' annual camp, started with a few heterogeneous units (field, works, bomb disposal, resources, survey, transportation and postal) but expanded considerably, having proved their value in the Suez campaign of 1956.

As the shape of the regular Army continually developed in the last half of the 20th century in line with changing defence commitments, so the role and scale of reservists were kept under scrutiny and greatly modified. Radical change overcame the Territorial Army in 1967 when the divisional organisation was abandoned and it was replaced by the Territorial and Army Volunteer Reserve, absorbing the Army Emergency Reserve, with a role confined to little more than providing reinforcements and specialist back-up to the regular Army. Later the pendulum swung back in favour of an increased profile for an organisation that was seen as providing good value. For the sappers there followed a period when the Territorial regiments had well-defined war roles in BAOR, until the collapse of the Warsaw Pact led to further reviews. By the turn of the century the engineer order of battle had been reduced to five regiments and a number of specialist pools and teams. In this form the Corps element of the 'citizen army' seemed to have found its true calling by providing fourteen per cent of the total Royal Engineers deployed on the first phases of the 2003 Iraq War (see page 222).

SOME HISTORIC RESERVE ARMY BADGES

Busby grenade,
1st City of London
Engineers

Cap badge,
Royal Jersey
Militia

Cap badge,
London Engineer
Volunteers

Glengarry badge,
Scottish CTC (Cable
Telegraph Company)

The Royal Engineers' association with the General Post Office began in 1870 when the latter became responsible for the provision of telegraph services. At that time the Corps provided telegraphic communications for the Army (see page 115) and through its association with the General Post Office its communication provision was eventually extended to include a postal service by taking over the duties of the Army Post Office Corps and forming the RE (Postal Section) in March 1913.

The postal arrangements for the Army had originally been provided by the civil Postmaster General, an Army Postmaster being appointed in 1799 to accompany the Duke of York's expedition to Holland. In the Peninsula a senior NCO was allocated to work with the Portuguese Post Office to supervise the military mail until 1811 when Major George Scovell of the Corps of Mounted Guides was put in charge of all communications. Similar exiguous arrangements existed for the Crimean War.

The Army Post Office Corps had been raised in 1882 on the eve of war with Egypt from the 24th (late 49th) Middlesex Rifle Volunteers to accompany the expedition (recruited entirely from GPO staff). Its function had earlier been agreed after a proposal by Lieutenant Colonel John Lowther du Plat Taylor, a previous Commanding Officer, but had not been implemented because of the expense.

The arrangements continued for the Boer War but despite unstinting efforts they proved insufficient for the task. By 1914 a Director of Postal Services had been appointed. A Reserve and Territorial organisation associated with the Royal Engineers had been set up and ten officers and 270 soldiers went straight out to join the BEF. It took time before the operational or logistic supply systems settled down to allow for the postal service to reach the necessary efficiency. However, as early as April 1915, Sir John French was able to report favourably that a letter posted in London was being delivered to GHQ on the following evening and reached an addressee in the trenches on the second day of posting.

These early beginnings brought together the essential characteristics of a military postal service: its foundation in the civilian Post Office structure, its integration into the military system and the fundamental difference from the normal supply chain, namely that mail destined for a named individual cannot be treated as a non-specific commodity such as rations or ammunition. Thus the requirement was for an efficient Home Depot in close touch with both the Government ministries and the Post Office, Base post offices in the theatre of operations and forward field post offices at formation headquarters. In between there has to be reliable transport, the acquisition of which would frequently stretch the creative talents with which postal staff were often so richly endowed.

Above: Letter From Home, the statuette commissioned by the Corps in 1993 with the help of donations from officers of the Postal and Courier Service to mark their transfer to the Royal Logistic Corps. The piece is a one-ninth scale replica in silver of the statue by Charles Jagger erected at Paddington Main Line station in 1922 as the Great Western Railway war memorial. It symbolises everything that mail means to the morale of soldiers in the field, a fact recognised by commanders throughout history.

THE HOME DEPOT

In 1914, the Army Post Office was set up initially at the GPO building at Mount Pleasant, letters and parcels going separately to different locations in London. It soon moved, eventually settling in Regent's Park. Later in the war some of the work had to be diverted to other main post offices elsewhere in the country.

Above: A Field Post Office in Natal. Army Post Office Corps (APOC) personnel were recruited entirely from staff of the General Post Office as members of the M Company 24th Middlesex Rifle Volunteers. In 1913 the APOC was re-organised to form the Royal Engineers Special Reserve (Postal Section)

Above: First Air Mail, painted by Terence Cuneo in 1978, commemorating the first scheduled airmail flight from England (Folkestone) to Germany (Cologne) on 1 March 1919. RE (Postal Section) are shown unloading an RAF DH 9 aircraft with the twin towers of Cologne Cathedral in the background. Thus could the RE (Postal Section) claim to have blazed the trail for the now universal delivery of mail by air.

Right: Field Post Office, Burma 1943, drawn by Major G. Dennison, RE, who served as a Deputy Assistant Director Army Postal Services during the Burma Campaign of 1943–5. Mail for the forward units fighting the Japanese in the recapture of Burma was airdropped and distributed to units by the formation field post offices. Due to the prevalence of disease among British troops, the British field post offices were often largely manned by Indian staff of the Indian Army Post Office.

Similarly in the Second World War initial mobilisation took place at Mount Pleasant but for most of the war it was at Nottingham. The scale of the work there is illustrated by the figures for 1944 when 3,000 men and women despatched 340 million letters, 95 million packets and newspapers and 13 million parcels in two-and-a-half million bags to troops overseas.

After the war the Depot was on the move again until it finally settled at Mill Hill.

EVOLUTION

After 1945 Britain's forces were still spread widely round the globe and actively engaged in post-war conflicts. Gradually commanders and the Ministry of Defence began to see the need for a regular structure backed by the Territorial Army and a closer working relationship throughout the chain of command with postal units as part of the normal military structure.

Integration between the services was the next logical step; BAPOs became BFPOs, an early trend towards the rationalisation that was to sweep the Services more generally decades later. Diplomatic mail to many parts of the

world was added. It was harder to win the case for taking over courier responsibilities from the Royal Signals, but the sense of this was accepted at the end of the 1950s and the REPS became the RE Postal and Courier Service. This step finally confirmed the need for RE (PCS) as part of the Regular Army. By 1979 this had been given effect in the form of four RE (PCS) regiments able to provide an RE PC squadron to support each division. This remained the shape of the service until the Corps finally said a sad goodbye to its foster-child on the formation of the Royal Logistic Corps in 1993.

Right: Date stamps of the Forces Postal Service from Hong Kong and Berlin in the 1990s. Date stamps, originated in 1660 and used to cancel mail as proof of posting, are the postal services 'tools of the trade'.

Above: Delivery of mail to the Royal Navy during the Falklands War

Left: A scene in the Home Postal Centre RE, Nottingham, in May 1943. Women of the Auxiliary Territorial Service are redirecting casualty mail. By the end of the war women made up some forty-nine per cent of the strength of the Home Postal Centre. The Royal Engineer grenade can be seen on the left breast of their uniform, a practice that continued for women attached to sapper units until 1992 when those of the Women's Royal Army Corps, who were trained as Postal and Courier operators, became members of the Corps.

Above: Air mail form ('bluey') (left) and airgraph (right). From the earliest days of air mail, methods were sought to reduce the payload. The airgraph was devised to reduce the bulk of mail from overseas during the Second World War. Servicemen wrote their letters on special forms. These were then reduced photographically, despatched as film then printed and sent to the recipient shown.

VI
NEW WORLD ORDER

39 | KOREA

THE EARLY BATTLES

As the icy Korean winter tightened its grip on the country in December 1950, 55 Independent Field Squadron deployed with 29th British Brigade just south of the North Korean capital, Pyongyang. It was the first British formed sapper unit to arrive in the theatre. The brigade was part of the rearguard for the withdrawal of the American and South Korean troops following the Chinese intervention in support of North Korea. It had been specially prepared in UK, the only other British force in the country having been two infantry battalions sent hurriedly from Hong Kong and, after being joined by an

Right: The Korean War Centre-piece is a model of the Great Gate of Seoul. It was bought from Mess funds to commemorate the war and carries a list of all engineer units that took part.

Below and right: Propaganda leaflets, a safe pass for deserters and a soldier's 'last letter home'.

Left: The Hook, by Terence Cuneo, depicts men of 55 Field Squadron repairing the defences one night in April/May 1953 after a period of heavy shelling. The Hook was a hill feature on the left of 1st Commonwealth Division's front, crucial to the defence. 1st Battalion the Black Watch had withstood a determined Chinese attack there in November 1952. On their return to the area in April 1953 the defences had to be strengthened to allow for artillery fire to be brought down on the position without withdrawing the defenders. In the picture is the troop commander, Captain (later General Sir George) Cooper, with a Black Watch officer. The framework of the cut-and-cover trenches and bunker, including reinforced concrete beams, can be seen. Soon after the period portrayed, the Chinese increased the intensity of their onslaught on the Hook. The Duke of Wellington's Regiment had taken over the sector on 13 May and two weeks later repulsed an attack that had been preceded by an artillery concentration of some 20,000 rounds, after fierce hand-to-fighting. Above the painting is a photograph taken after a Chinese attack.

Right: A map recording minefields in the 1st Commonwealth Division area up to 21 May 1953.

Australian battalion, designated 27th Commonwealth Brigade.

By that time the war had been in progress for six hectic months. North Korea, a Soviet Russian client state since the end of the Second World War, had without warning invaded the American-controlled south in June 1950, thus frustrating the unification of Korea that had been approved by the United Nations in 1947.

With the approval of the United Nations Security Council, which by chance the Soviet Union was currently boycotting, the Americans sent forces to support the South Koreans. They held on to a small perimeter round the southern port of Pusan until the United Nations force had built up enough strength under General Douglas MacArthur to launch a surprise landing on the west coast at Inchon. By October the North Koreans had been defeated and pushed back close to the Chinese border. The Chinese then

joined the battle and the United Nations were forced back, first to about the 38th Parallel and, after a second offensive, to a line south of Seoul where they held firm.

During this withdrawal 55 Squadron established a fine reputation with the Americans,

Right: The Korean War memorial, in the Garrison Church, Brompton, was unveiled on 24 April 1988 by the Chief Royal Engineer, General Sir George Cooper, GCB, MC. It was the work of the sculptor John Skelton.

contributed to by two particular achievements. In the first they built a Class 30 floating bridge to support the withdrawal across the mighty Han river using American M4 equipment, which they had never seen before, and succeeding in completing at US Engineer text-book rates. In the second they demolished the 66-span Han road and rail bridge in exceptionally harsh conditions (see page 200). This was one of many demolitions prepared by the Squadron during the withdrawal in which, among their other tasks, were road work on the scarce and inadequate routes and helping the infantry with wiring and mines; all this in the severe weather for which no special winter clothing had yet been provided.

This period of operations was followed by an advance back to the Imjin line, to which the Chinese responded with their massive spring offensive (in which the 1st Battalion the Gloucestershire Regiment was cut off) and a further

Above: The water point at Pintail bridge on the Imjin. The tanks were contained in heated storage to keep the water from freezing in the extremely cold winter.

fighting withdrawal by 29th Brigade. This, however, succeeded in defeating the Chinese attempt to break through to Seoul. At the same time the British contribution was strengthened by the creation of 1st Commonwealth Division, now including a Canadian brigade. 27th British Brigade was redesignated 28th Commonwealth Brigade. The field engineer component of the Division was then 28 Field Engineer Regiment (12 and 55 Field Squadrons with 57 Independent Field Squadron RCE) and 64 Field Park Squadron, all under a CRE, the first of whom was Colonel E. C. W. Myers.

CONSOLIDATION

By the end of May 1951 1st Commonwealth Division had taken up positions along the line of the River Imjin. The pattern of operations was vigorous patrolling of the 7km-deep sector across the river within which lay a Chinese light outpost screen in front of their main line. Many of these operations developed into offensive raids supported by tanks. The river crossings, ferries established at two crossings known as Pintail and Teal, were a major concern for the divisional sappers, particularly as the river rose dramatically in the rains.

Ceasefire talks began in July 1951 but it was clear that the Chinese were only interested in the opportunity these gave to strengthen their positions. In the autumn it was decided that the United Nations line in this sector should be pushed forward to the higher ground some ten to thirty kilometres to the north of the river. A major attack was launched across the front in conjunction with the flanking divisions, in which three detachments of sappers advanced with the divisional RAC regiment to check likely places for mines. After bitter fighting in some areas all objectives had been taken by 8 October. It was some time before these positions were firm and properly dug in after a huge programme of minelaying and building field defences that could withstand the intense bombardments of the prolific Chinese artillery. Nevertheless a determined series of attacks in November was kept at bay and gradually the positions grew stronger. Apart from one small redeployment on the east of the line, the divisional area now remained unchanged throughout the rest of the war.

Establishing the new line called for enormous sapper effort, but much of that had to be expended on a heavy minelaying programme and on simply maintaining communications in the extreme winter conditions, including building new jeep tracks in the forward areas. Sapper strength was augmented by the invaluable Korean Service Corps, a quasi-military organisation of men unfit for the Korean Army but who nevertheless performed well, often under fire, suffering many casualties. Throughout 1952 the pattern of operations in this now static line was one of vigorous patrolling and a gradual strengthening of the forward posi-

tions with field defences, including bunkers for command posts and gunner OPs, cut-and-cover trenches and tunnelling. Forestry camps were established to provide for the huge quantities of timber now demanded. Mine warfare became progressively more significant and special training was conducted within the division in an effort to reduce casualties from both enemy and our own minefields.

After a period in reserve during the winter, the Commonwealth division moved back into the line in the first week of April 1953. By that time they had inherited from the United States Marine Corps the key feature known as the Hook. Measures to strengthen it continued apace (see page 196). The final major (as it turned out) attack on the Hook was made on 28 May 1953.

CEASEFIRE

The peace negotiations that had dragged on for more than two years finally bore fruit in the form of a truce on 27 July 1953. A Demilitarised Zone was set up, which exists to this day. There was an immediate withdrawal to south of the Imjin where the Commonwealth Division developed their old position on the 'Kansas Line' in case the truce were to break down. As the likelihood of this receded, the Commonwealth Division disbanded. 28 Engineer Regiment, less 55 Field Squadron, returned to the United Kingdom in March 1955. In May 1956 55 Field Squadron rejoined them for the Regiment's next assignment on Christmas Island, having completed five and a half years in Korea.

Top left: Road camouflage. Many of the roads in the forward areas were under direct enemy observation and frequent targets for artillery.

Centre left: The 1,000-foot Han River 'Shoofly' bridge demolished by 55 Independent Field Squadron in January 1951 in support of the United Nations' withdrawal. The bridge was in continuous use during the preparation. At one point a major disaster was averted when a sapper spotted a cutting charge which had been ignited by a spark from a passing train and kicked it into the river. Each pile of the 66-span bridge was cut below the water-line. The demolition was a complete success.

Bottom left: Attack on a village, March 1951. Tanks of the 8th Hussars and infantry of the 1st Glosters are fanning out across the fields. Sapper carriers with mine lifting parties are behind the tank. One sapper is walking in front of the tank.

The British nuclear test programme was launched in the Montebello Islands, sixty miles off the north coast of Australia in October 1952. Over the next six years there was a series of tests in Australia, on Malden Island (400 miles south-west of Christmas Island) and Christmas Island itself (see page 203 for full list). Although the Corps was represented at all of these tests, it was the early ones on the Montebellos and the final series on Christmas Island itself that led to the greatest commitment.

The Montebello tests were organised under naval command. For the first, 180 Engineer Regiment was raised on a specially designed establishment. After six months' preparation and planning the Regiment arrived in the Montebellos in April 1952. The main tasks were base building, reinforced concrete pillboxes

The British nuclear test programme, in all of which the Corps gave detailed support, ran from 1952 to 1958, culminating in the megaton tests on Christmas Island.

for cameras and recording equipment, several towers for cameras and aerials, Nissen huts for generators, stores, workshops and messes and about fifteen miles of light roadway. Construction took seven months, and the tests were completed by October. Much of the accommodation was aboard naval vessels and Landing Ships, Tank. For the second group of tests in 1956, an augmented field troop only was needed to reinstate the necessary facilities, a four-month task followed by the tests in June.

By this time the development of Britain's hydrogen bomb was well advanced. Tests to take place on Christmas Island in the Pacific Ocean were urgently required. Lieutenant Colonel (later Major General) John Woollett, who had just brought 28 Engineer Regiment less 55 Field Squadron back from Korea, was appointed CRE in January 1956. Intense planning took place and by July the whole regiment was on the island including 55 Squadron who came straight from Korea. Their tasks included the repair of existing port facilities, shore accommoda-

tion including special buildings for the Atomic Warfare Research Establishment work, and the reconstruction of the runway.

While this, and the successful firing of Britain's first thermonuclear weapon, was in progress, the decision had been taken to set up a permanent test base on the island. Massive expansion of the facilities was called for and a Joint Task

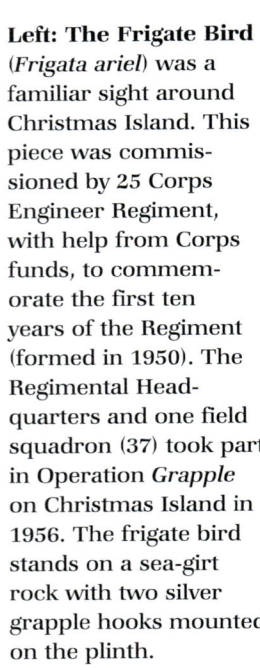

Left: The Frigate Bird (*Frigata ariel*) was a familiar sight around Christmas Island. This piece was commissioned by 25 Corps Engineer Regiment, with help from Corps funds, to commemorate the first ten years of the Regiment (formed in 1950). The Regimental Headquarters and one field squadron (37) took part in Operation *Grapple* on Christmas Island in 1956. The frigate bird stands on a sea-girt rock with two silver grapple hooks mounted on the plinth.

Left: The Starmix 40 asphalt site was critical to the progress of the construction of the runways of the two airfields built on the island. The plant was set up in January 1958, replacing three smaller machines. A party of an officer and twenty-eight men was needed to maintain two-shift working together with twenty-six men operating the bitumen heating tank farm. Aggregate was supplied from four Parker rock crushers, which also supplied the almost insatiable demand for concrete.

UK NUCLEAR TESTS 1952–1958				
Codename	Location	Date	Yield Range	Explosion Conditions
Hurricane	Montebellos (off Trimouille Island)	3 Oct 1952	25 kilotons	Ocean surface burst (HMS *Plym*)
Totem 1	Emu (Maralinga)	15 Oct 1953	10 kilotons	Tower mounted
Totem 2	Emu (Maralinga)	27 Oct 1953	8 kilotons	Tower mounted
Mosaic G1	Montebellos (Trimouille Island)	16 May 1956	15 kilotons	Tower mounted
Mosaic G2	Montebellos (Alpha Island)	19 Jun 1956	60 kilotons	Tower mounted
Buffalo	Maralinga (One Tree)	27 Sep 1956	15 kilotons	Tower mounted
Buffalo	Maralinga (Marcoo)	4 Oct 1956	1.5 kilotons	Ground burst
Buffalo	Maralinga (Kite)	11 Oct 1956	3 kilotons	Air dropped – high air burst over land
Buffalo	Maralinga (Breakaway)	22 Oct 1956	10 kilotons	Tower mounted
Grapple 1	Malden Island, Pacific	15 May 1957	Megaton	Air dropped – high air burst over ocean
Grapple 2	Malden Island, Pacific	31 May 1957	Megaton	Air dropped – high air burst over ocean
Grapple 3	Malden Island, Pacific	19 Jun 1957	Megaton	Air dropped – high air burst over ocean
Antler	Maralinga (Tadje)	14 Sep 1957	1 kiloton	Tower mounted
Antler	Maralinga (Biak)	25 Sep 1957	6 kilotons	Tower mounted
Antler	Maralinga (Taranaki)	9 Oct 1957	25 kilotons	Balloon suspended – high air burst over land
Grapple X	Christmas Island	8 Nov 1957	Megaton	Air dropped – high air burst over ocean
Grapple Y	Christmas Island	28 Apr 1958	Megaton	Air dropped – high air burst over ocean
Grapple Z	Christmas Island	22 Aug 1958	Kiloton	Balloon suspended – air burst over land
Grapple Z	Christmas Island	2 Sep 1958	Megaton	Air dropped – high air burst over ocean
Grapple Z	Christmas Island	11 Sep 1958	Megaton	Air dropped – high air burst over ocean
Grapple Z	Christmas Island	23 Sep 1958	Kiloton	Balloon suspended – air burst over land

Below: The main camp on Christmas Island. The tents of the 1956 deployment were gradually replaced by prefabricated huts. By 1958 accommodation for 3,000 all ranks had to be allowed for. Priority was given to communal facilities such as messes and for administrative units: the hospital, workshops, laundry and bakery. Emphasis was placed on recreational provision with asphalted hockey pitches and tennis and basketball courts.

Force Headquarters set up under an Air Vice Marshal with a full Colonel as Chief Engineer. Three engineer regiments (25 Field and 38 and 36 Corps) and numerous independent field squadrons followed 28 Regiment, to build up the huge complex and support the tests during the period from March 1957 to September 1958. At its peak the sapper workforce rose to over 1,800. The last regimental tour continued to November 1959 to carry out works consequent upon the decision thereafter to end weapon testing in the atmosphere. The specially established 73 (Christmas Island) Squadron then remained to look after the facilities until the island was fully restored to its Gilbertese inhabitants in 1964.

Christmas Island was the largest single project undertaken by the Corps in the post-war years. It demonstrated the versatility of regiments trained essentially for field operations but with a hard-core of tradesmen and technicians, to turn their hand at remarkably short notice to works on this scale. It also gave those who took part an insight into the realities of the war for which the armed forces now had to train and organise. As the Chief Engineer during the final tests, Brigadier Robert Muir, put it: 'On reflection perhaps the most outstanding personal impression from nuclear weapon explosions was the reality of the horrific overkill propensity.'

It is my duty, however, to place before you certain facts about the present position in Europe. From Stettin in the Baltic to Trieste in the Adriatic an iron curtain has descended across the Continent.

Winston Churchill, Fulton, Missouri
5 March 1946

These words heralded the Cold War, the ideological confrontation between the 'Eastern bloc' and 'Western Bloc' states led by the Soviet Union and the United States respectively, formalised by the North Atlantic Treaty Organisation alliance in 1949 and the Warsaw Pact in 1955. This confrontation transformed the British Army of the Rhine from a post-war occupation force into a field army ready to deploy, as part of the alliance, against what was seen as an imminent threat of invasion from the east. The NATO commitment became Britain's top defence priority involving not only the troops stationed in Germany but also

Left: General Sir Charles 'Splosh' Jones, GCB, CBE, MC, 'fearless fighting sapper' as The Times called him, was endowed with exceptional qualities of leadership and 'a rare talent for war'. His outgoing personality and professional dedication took him from an early career with the Bombay Sappers and Miners through CRE 79th armoured division to post-war command of 2nd Infantry Brigade, 7th Armoured Division, I (British) Corps and Northern Command. His active career ended as Master General of the Ordnance and Army Board member after which he became Governor of the Royal Hospital, Chelsea.

Left: Bridge demolition was a much-rehearsed art in the BAOR training cycle. Many of the bridges within the potential battle zone were pre-prepared with chambers and tailor-made charges stored in readiness. The German militia *Wehrbereichskommando* maintained the details of these and trained with engineer units on field exercises. This picture is a watercolour by Ken Howard, RA, whose work, capturing the mood of contemporary military training, was reproduced in limited editions, to familiar unit messes.

United Kingdom-based formations. All training and equipment within BAOR was focused on the single aim of fighting in Western Europe in a scenario, based on the best intelligence of the time, in which there would be little warning of such an attack. This demanded a high state of readiness to deploy to war and all the logistic and reinforcement back-up so implied. This stand-off lasted until the collapse of the Soviet Union led to the formal end of the Warsaw Pact in 1991.

Thus did all sappers in this period find themselves at one time or another serving in Germany either at one of the permanent stations or on exercise from the United Kingdom. As the Warsaw Pact threat developed in the 1950s, training for war was governed by the General Deployment Plan that emanated from the Supreme Allied Commander, Europe, through army group headquarters to the national corps in the forward areas. Under this plan, units would move out of barracks on receipt of coded orders. Sappers would expect to receive their orders ahead of the armoured divisions so as to be able to prepare the obstacle plan, which included laying minefields, preparing demolitions and other works to enable the divisions to move and fight in their areas. A carefully choreographed procedure whereby stores were outloaded and units moved to pre-planned areas was rehearsed every year in a series of exercises in areas of Germany in accordance with the rights for the use of land outside designated training areas established by agreement with the German government.

These arrangements did not begin until about 1949. By 1948 the post-war army in Germany had run down to a single division, (2nd Infantry Division). 23 Field Engineer Regiment, based at Hameln, was the only field engineer regiment in the country and provided the divisional engineers. In 1949 21 Engineer Regiment was formed as the divisional engineers for the newly created 7th Armoured Division (commanded by the sapper Major General Charles ['Splosh'] Jones). Tension increased with the outbreak of the Korean War

BERLIN
Berlin was the most manifest symbol of the Cold War. Divided into four sectors governed by American, British, French and Russian military commanders as agreed in the Yalta and Potsdam conferences, the city was used by the Russians to test the resolution of the Western allies to prevent this outpost of Western democracy being absorbed into communist East Germany. Their 1948 blockade of the city was frustrated by the Berlin Airlift that ran for a year before the border controls were eased. Sappers played a significant role in this success through the efficient movement control system, the maintenance of services and repairs and extensions to the runway at Gatow airfield.

Thereafter West Berlin was controlled separately and apart from the Russian zone. The Berlin Wall was built in 1962 but access to East Berlin remained available subject to minor controls.

Similar controls were imposed for overland travel across the 'Inner German Border', with its wire and minefields, by single corridors for road and rail through East Germany to Berlin, an attractive location for sporting and leisure trips from the 'zone' of north-west Germany.

The Corps was originally represented in Berlin mainly by a DCRE and field troop. In 1957 38 (Berlin) Field Squadron was established as support for the Berlin Brigade. An Area Works Officer (lieutenant colonel) controlled what was then the only permanent military works area.

Above: The Brandenburg Gate, a silver model crafted by Herr Albert Eggert for 38 (Berlin) Field Squadron.

ARMOURED ENGINEERS

Only in 1957 did assault engineers re-appear in Germany after their post-war return to the United Kingdom where the capability had been maintained in 32 Assault Engineer Regiment and 113 Assault Regiment (TA), both of which had been disbanded by 1957. Tentative trials with 26 Armoured Engineer Squadron, as it then became known, as part of an experimental armoured force, sowed the seed which eventually burgeoned as 32 Armoured Engineer Regiment in 1964. It was equipped with bridgelayers, AVREs and ARKs, all based on the Centurion tank that had replaced the Churchill in the 1950s. The armour in the regiment was then split to give the 1969 regiments an organic capability for their divisions. The restructuring proposals in 1975 excluded the AVRE in favour of equipping the regiments with the Combat Engineer tractor and leaving the bridgelayers the only armoured engineer equipments. Trials proved this to be unwise, and the AVRE was gradually restored, first to a revived 26 Armoured Engineer Squadron and, later

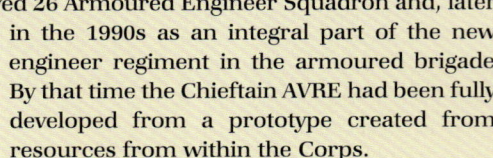

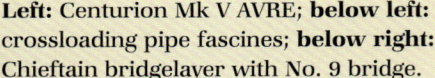

in the 1990s as an integral part of the new engineer regiment in the armoured brigade. By that time the Chieftain AVRE had been fully developed from a prototype created from resources from within the Corps.

Left: Centurion Mk V AVRE; **below left:** crossloading pipe fascines; **below right:** Chieftain bridgelayer with No. 9 bridge.

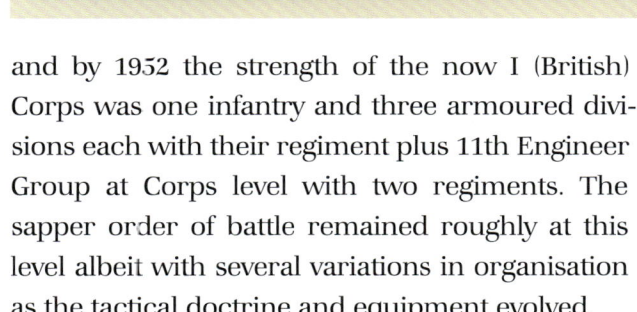

and by 1952 the strength of the now I (British) Corps was one infantry and three armoured divisions each with their regiment plus 11th Engineer Group at Corps level with two regiments. The sapper order of battle remained roughly at this level albeit with several variations in organisation as the tactical doctrine and equipment evolved.

The organisation of the division into regiments was a departure from wartime practice, but with the arrival of tactical nuclear weapons it was envisaged that brigades would have to be more widely dispersed around the battlefield than hitherto and that they must have their dedicated field squadron in support. Commanding officers became Commanders Royal Engineers, part of

Left: The Haynes Medal commemorates Captain Alfred Haynes who was killed in 1896 in Matabeleland (in today's Zimbabwe). He had volunteered to lead a party of some forty Royal Engineers to accompany a force ordered to put down an insurrection under Chief Makoni. Haynes was killed in the successful assault on Makoni's kraal. A fund was raised by his brother officers for a memorial in Rochester Cathedral and to endow a medal to be awarded to a sapper in each party of recruits on the fieldworks course at the School of Military Engineering. It is now awarded to the student who passes out top of the Junior NCOs' Instructors' course.

Left: The vertical take off and landing Hawker Siddeley Harrier, introduced into service in the early 1960s, brought a specialist task to the Corps: the reconnaissance, construction and concealment of forward operating sites. Initially this support was provided solely from the United Kingdom, but as techniques developed it became clear that specialists in the theatre were needed and 10 Field Squadron was formed and allocated both to this role and to the equally critical one of rapid runway repair on the permanent airfields.

Left: The Barmine-layer entered service in the late 1960s to provide for the rapid laying of the tens of thousands of mines demanded by the General Deployment Plan. This example at the RE Museum saw subsequent service in the Gulf War.

Above: The Freedom of Nienburg was granted to the Corps on 27 July 1980. The British custom of granting the freedom of a town to a regiment, thereby permitting them to march through its streets with bayonets fixed, drums beating and flags flying was adopted by Hameln, Osnabrück and Nienburg, the three German stadts that had been hosts to sapper units longer than any other. In all cases the Freedom was granted to the Corps rather than to the actual regiments.

BRIDGES AND FERRIES IN THE POST-WAR YEARS

Bridge and ferry development had to keep pace with the rapidly increasing loads, particularly tanks, that came into service to meet the Warsaw Pact threat. At the same time construction times had to be reduced to match increasing battlefield mobility. The Heavy Girder Bridge was a development on Bailey but too heavy for use in forward areas. Ferries and floating bridges carried on dedicated transport, such as the Heavy Ferry, Heavy and Light Floating Bridges, from which the elements could be directly launched, filled the need for many years. The problem of rapid crossing of wide rivers in realistic operational conditions was not really solved until the arrival of amphibious equipment in the early 1960s. The French *Gillois* with inflatable floats was later replaced by the German M2B and then the M3, an Anglo-German collaborative project. The demand for a rapid construction dry bridge was met by the highly successful Medium Girder Bridge, which entered service in 1971, bringing with it a new dimension of simplicity of construction and versatility.

Throughout the 1980s an advanced system of bridging was introduced, in collaboration with allies, whereby a minimum of common parts could be used for both close and general support applications.

In the end the new BR90 system emerged with its purpose-built Automotive Bridge Launching Equipment, bringing a further step change in versatility and speed of construction.

The Heavy Assault Floating Bridge (top left) was the principal floating bridge for the British Army of the Rhine in the 1960s and 70s, seen here on a Class 100 trial carrying a Conqueror tank on its transporter.

The Medium Girder Bridge (top right), seen here on deployment in Bosnia, came into service in the 1970s to provide a hand-built Class 60 crossing. It proved highly versatile and was successfully sold around the world.

The M2 Amphibious Bridge (bottom right), a German equipment, came into service in 1962, replacing the French *Gillois* which depended on inflatable pneumatic floats. In the main illustration, the rig is being prepared for launching and subsequent joining up as a ferry or complete bridge (a painting by Ken Howard).

BR90 (Bridging for the 90s) (bottom left) provides a highly versatile system that can provide both close and general support bridges capable of spanning gaps from nine to sixteen metres. All bridges can be built from seven modular panels.

The Heavy Ferry (centre left) was developed in the post-war years to provide a Class 80 bow-loading equipment capable of being built in an hour to provide for the immediate support of an assault crossing with armour and support weapons. (A painting by John Young.)

the divisional staff with a field park squadron at divisional level together with whatever allocation of corps troops that could be made available. By 1965 the engineer regiments had re-formed, their commanding officers doubling as CREs. In 1969 the corps engineer regiments were disbanded to provide the manpower for two regiments in each division answerable to a full Colonel CRE at division. This arrangement lasted until 1978 when a radical restructuring of I (British) Corps envisaged the abandonment of the brigade level of command. Although trials soon proved this unworkable, divisional regiments came back into favour, the divisional Commander Royal Engineers being retained.

The apparently relentless growth of the Warsaw Pact forces with their proliferation of armour and capability for nuclear, biological and chemical warfare led to full mechanisation of the forward divisions in BAOR in the 1960s. Numerous specialist tasks faced the Corps to allow the army to live and fight in an environment yet to be experienced in warfare. Operational concepts envisaged a bleak scenario with massive destruction of forests and centres of population, widespread nuclear and chemical contamination of the battlefield and a proliferation of air-delivered scatterable mines. Preparedness for this catastrophic situation was all part of the policy of deterrence that eventually led to the collapse of the Warsaw Pact in 1991 and a reassessment of defence policy within the Atlantic Alliance.

Following that event, yet more reorganisation gave the opportunity to readjust the sapper order of battle so that sufficient armour was integrated into the regiments to enable them adequately to support their divisions. The Corps eventually emerged from the defence reviews of the 1990s with a regiment for each armoured brigade equipped with its own AVREs and Chieftain bridge-layers. This organisation proved its value in the difficult circumstances of peace-keeping in the Balkans as well as in the more mobile actions of the wars in the Gulf and Iraq (see page 221).

Above: The Combat Engineer Tractor was introduced in 1978. It was the first such machine designed from scratch for engineer use after the Second World War. Capable of performing tasks in close support of mechanised troops, it had a 1.7 cubic metre bucket, was fully amphibious, could carry a rocket-propelled earth anchor and had fittings for towing Giant Viper and laying trackway.

42 | WINDS OF CHANGE

Above: 1918–62 GSM 'Canal Zone', awarded only in 2003 after prolonged lobbying by veterans who had fought in this war zone.

The wind of change is blowing through the continent. Whether we like it or not, this growth of national consciousness is a political fact.

Harold Macmillan, 3 February 1960
Houses of Parliament, Cape Town

While the British Army of the Rhine was training for possible nuclear war in Europe, elsewhere it was engaged in a series of actual conflicts in the process of handing over power in the former British dependencies. Britain's responsibilities as the sovereign power and arbitrator between local, often violent, factional interests demanded a completely different set of skills from those of the Cold War. Techniques in counter-insurgency, counter-terrorism and simply 'keeping the peace' became the speciality of the UK-based army within which a strategic reserve had to be available to move at short notice to back-up stationed forces in the area of conflict.

THE MIDDLE EAST

The first of these was in Palestine, a British responsibility under mandate from the League of Nations and a candidate at that time for a possible strategic base after the planned evacuation of Egypt. The Jewish militants were well organised and able to exert such pressure for the establishment of a homeland on their terms, incompatible with the aspirations of the Arab inhabitants, that in the end, following a United Nations decision for the partition of the country, the British government ended the mandate and withdrew in May 1948. In the process three British Army divisions had had to deal with relentless civil disturbance, terrorism and sabotage. For the sappers, coping with mines and booby traps, maintenance of public utilities and works connected with the post-war reorganisation were the main tasks. The most notorious event of the period was the blowing up of Jerusalem's King David Hotel in July 1946 in which ninety-six people were killed. The rescue work and rubble clearance was carried out with much credit by 9 Airborne Squadron. The eventual withdrawal was a masterly affair in which engineer work in every field including Transportation, Works and Stores were at full stretch. As the last troops left, the 1948 Arab-Israeli War began. Twenty-two Royal Engineers had lost

Right: A communist guerrilla's hat from the Malayan emergency.

Left: Gazelle Bridge, Bukit Mendi, Malaya, exemplifies a project under the Military Aid to the Civil Community scheme designed to give experience to sapper officers, technicians and tradesmen and so also benefiting the local population. The bridge was built between December 1968 and January 1970 under an agreement between the Army and the Federal Land Development Authority. Design and control was by 63 CRE (Construction) and the work was executed by troops found in turn by 11 and 59 Field Squadrons and 67 Gurkha Field Squadron, with detachments from 54 Support Squadron, whose soldiers are depicted here. (Artist Norman Hepple.)

Far left: **A company base during the Borneo Confront-ation**. Most of these bases were accessible only by air or river. Sapper support included the construction of helicopter pads, command posts and defence and camp structures, and the provision and maintenance of water supplies.

Near left: **Meligan airstrip** in the Fifth Division of Sarawak, built with plant 'knocked down' for carriage in under-slung helicopter loads and reassembled on site.

their lives in three years of peacekeeping duties.

British power in the Middle East was then centred on the Canal Zone in Egypt to which the Army Headquarters had withdrawn from Cairo and in which the vast stocks of wartime stores had accumulated. The sapper presence was considerable: the Chief Engineer presided over two CREs (Works), an engineer stores base depot, plant depot and workshop, the Middle East School of Military Engineering, the Survey Directorate, the Transportation and Movements Directorate including 4 Port Operating Group and 10 Railway Squadron, and an army engineer regiment.

Reorganisation in the theatre led to a colossal project to set up a stores base at Mackinnon in Kenya, begun in 1947. By 1950 this new base with 80,000 tons of stores and supporting elements including a 600-bed hospital, was a going concern. Almost at once the strategic situation changed and the project had to be abandoned.

In 1951 the Egyptian government abrogated the

Below: **Habilayn airfield** was the forward base for operations in the Radfan. This picture (*The Flying Postman*) by Terence Cuneo commemorates the activities of the Royal Engineers Postal and Courier Service, who participated in the campaign, and it was handed over to the Headquarter Mess in 1993 on the transfer of the Service to the Royal Logistic Corps in 1993. A Beaver aircraft is being unloaded of its mail. From Habilayn, mail would then be taken forward to isolated detachments up country where the bags would be dropped from a helicopter.

Right: A Z Craft on one of the many voyages made to transfer essential stores from the Canal Zone to Cyprus in 1955. (IWM HU51727)

Far right: A Ramp Powered Lighter off Aden. This type of vessel took over the role of the Z Craft.

Right: The Dhala Road was a thirty-six-mile stretch of road leading through Habilayn and fifteen miles on, and six thousand feet up to the mountain emirate of Dhala.

1936 Anglo-Egyptian Treaty. The internal security situation became acute with consequential massive build-up of troops. Works, camp structures and the maintenance of utilities absorbed huge sapper effort, to say nothing of the support of two infantry divisions and a parachute brigade. The Canal Zone was finally evacuated in 1956 in favour of the new base in Cyprus, where the Army was still fully engaged with the EOKA revolt that only came to an end two years later and had cost four sapper lives.

Immediately after the withdrawal the Egyptian President Nasser nationalised the Suez Canal Company, provoking the Anglo-French invasion of Suez. The British contribution to this was produced largely from BAOR, comprising 3rd Infantry Division, 3rd Commando and 16th Parachute Brigade; by Royal Proclamation, some 25,000 reservists were mobilised. In late October 1956, Israel attacked across the Suez Canal. The Anglo-French forces invaded on the pretext of parting the belligerents. Nasser refused to accept the presence of foreign troops in Egypt and out of the chaotic international diplomatic situation that arose the Allies were obliged to call the enterprise off and allow a United Nations observer force in to hold the ring. This episode, though politically ill-conceived, was efficiently carried out. The airborne assault on Gamil airfield Port Said on 5 November was followed up the next day by a commando brigade landing accompanied by 3 Field Squadron with 37 Corps Engineer Regiment in the next wave. The ceasefire was ordered at midnight on 6/7 November by which time the break-out had advanced twenty-three miles to the south. After a period of negotiations and repairs to the public

Right: Crown airfield, Thailand, built in 1964–7 as part of a South East Asia Treaty Organisation scheme for forward strategic airfields against the perceived threat of communist insurgency across the Laos border, close to which it was located.

Right: 37 Corps Engineer Regiment on anti-EOKA operations in Cyprus. The Cyprus emergency, resulting from the conflict between the Greek community who sought union with Greece (*Enosis*) and the Turkish minority who wanted partition, ran from 1955 to 1959. Sapper support included providing emergency deployment camps for reinforcement units.

utilities, the withdrawal was complete three days before Christmas.

President Nasser stretched British exasperation even further when in 1962 he intervened in the border dispute between Yemen and Aden, a British colony whose independence was then under negotiation. Thus began five years of one of the most intractable of Britain's post-war struggles to withdraw with some honour and if possible to bequeath a respectable economy and administration in exchange for some military facilities. Camps were already under construction to this end, but the situation deteriorated so quickly that expansion plans for reinforcement became essential. By 1964 the sapper presence of an independent field troop had grown to two field squadrons and from then on such units as were required over and above the stationed troops were provided by *roulement* from the strategic reserve. The increasing activity of the insurgents in the Radfan border area led to the need for the forward base at Habilayn and the Dhala road (see

Above: Ubique 1945–1988, the fine piece of silver commissioned from Mr Stuart Devlin to represent the countries and campaigns, not otherwise commemorated in silver, in which the Corps was involved between 1945 and 1988. The globe, mounted in gimbals, carries the outline of the countries concerned fitted together jig-saw style rather than geographically.

Centre right: A Caterpillar D4 tractor in airdrop mode on a medium stressed platform, the standard means of delivery from the Beverley aircraft, used throughout the Borneo campaign. (RE Museum exhibit)

Bottom right: Suez 1956, rubble clearance.

Above: Malta Sapper, an oil painting by Garth Legge, commemorates the long association of the Corps with Malta. It depicts a country scene with the cathedral of Musta in the background and Maltese sappers shaping blocks of stone. After the Second World War the Royal Engineer presence comprised the Malta Fortress Squadron and a CRE (Works), later reducing to a specialist team until its final withdrawal in 1977.

illustrations). Aden was evacuated in November amid scenes of increasing acrimony from the local population. The British presence in the area then continued through deployments to the Gulf and, in the 1970s, by support for the Sultan of Oman in his fight against attempted cross-border insurgency in Dhofar.

THE FAR EAST

Even while these convulsions were in progress, parallel events were engaging the army in the Far East. The post-war run-down had left only the Singapore Base with a largely Works establishment and the Singapore Engineer Regiment. By 1948 there was no engineer unit left on the Malayan peninsula. In that year there broke out the emergency created by the jungle-based Chinese communist guerrilla movement. Their attempts to subvert the workers on the rubber estates were defeated by the policy of resettlement into controlled villages combined with vigorous fighting patrols and ambush work deep in the jungles, which became the staple business of the infantry. Malayan independence was granted in 1957, but the end of the emergency was not officially declared until 1960; by that time the idea of a wider federation to include Sarawak and Sabah in North Borneo was under discussion and effected in 1963. Borneo then became the final theatre of counter-insurgency operations as Indonesia-based guerrillas first attempted to overthrow the Sultan's regime in Brunei and then infiltrated the border areas until this 'Confrontation', as it was known, was also defeated by 1966.

During the Malayan emergency the Gurkha Engineers were created (see page 174). At much the same time 76 'Federation Field Squadron' was raised on a nucleus of officers and soldiers of local origin who had served in Royal Engineer units, the seed of the future Malaysian Engineers. In the early days of the emergency sappers were in such short supply that the Engineer Training Centre at Kluang was frequently called upon to produce patrols and detachments for special tasks. At the height of the Borneo Confrontation there were two Gurkha field squadrons, a British field squadron on *roulement* from the UK and a field troop from Malaya, two Malaysian field squadrons and an Australian construction squadron in the theatre of operations.

The experiences of the three post-war decades earned the British Army a reputation for particular aptitude in counter-insurgency and peacekeeping. 'Winning the hearts and minds' of local populations, through simple projects that encouraged settlement and removed the causes of disaffection on which revolutionaries could feed, raised the profile of the Corps and gave it a leading role of its own.

Below: A twenty-mile track along the Mukeiras escarpment in Aden was built by the Royal Engineer Independent Troop in 1961. The photograph was taken from a Beaver aircraft.

Awards, right: Left, **General Service Medal 1918–62, Palestine**; centre, **General Service Medal 1962**, with Borneo and Malayan Peninsula clasps; right, the **Rhodesia Medal** awarded for service with the force sent at short notice in December 1979 to monitor the elections that followed the civil war in Rhodesia and form the basis of a new army by bringing together men who, albeit, had been fighting one another for many years. The sapper contingent was based on 22 Engineer Regiment with individuals drawn from five other units. The reaction of the twenty thousand guerrillas living in the bush was uncertain, but in the event they embraced the new Zimbabwe and the new army began to form up enthusiastically.

Right: A Royal Air Force Valetta drops supplies into Fort Kemar during the Malayan Emergency. This was the first of the jungle forts to be built with a view to offering protection to the local aboriginal tribes, who were being threatened by the communist terrorists. It was started by a party of an officer and ten Malayan sappers brought in by helicopter and using hand tools. It was completed using a Fordson tractor and towed equipments broken down into helicopter loads. In all, twelve of these jungle forts were built with airstrips capable of landing a Twin Pioneer aircraft. (IWM MAL 52)

Yomp – verb intransitive, Brit slang, march with heavy equipment over difficult terrain [1980s. Origin unknown.]

Oxford English Dictionary

THE WAR

On Friday 2 April 1982 Argentine forces invaded the Falkland Islands. This unprovoked act had been preceded by the planting of a party of so-called scrap merchants on South Georgia to which a party from the small Royal Marines detachment had been sent to maintain observation. The remainder of the detachment at Stanley put up a spirited resistance, but the Governor ordered them to lay down their arms once it was clear that the scale of the invading force was overwhelming.

Below: The airstrip at San Carlos, built by 11 Field Squadron as the Harrier forward operating base after the sinking of SS *Atlantic Conveyor* (see text). Fortunately sufficient trackway that had been separately loaded for other tasks was re-allocated. Also shown are the fuel supplies under construction with a Combat Engineer Tractor excavating sites for additional bulk fuel storage tanks. The three figures in the foreground illustrate the three elements of the Corps represented in the war, an officer from the mainstream and one each from the Airborne and Commando Sappers. (Painting by Terence Cuneo)
Left: an aerial photograph of Port San Carlos, where the landings took place.

Almost immediately the first elements of the task force planned to retake the islands set sail, led by the aircraft carriers HMS *Hermes* and HMS *Invincible* with 3rd Commando Brigade and the 3rd Battalion the Parachute Regiment, with their supporting arms including 59 Commando Field Squadron and a field troop of 9 Parachute Field Squadron. 11 Field Squadron also deployed early in the Harrier Support role. Against a background of intense political activity to obtain moral and material support from the United States, planning went ahead for what was to build up to a two-brigade operation by the addition of 5th Infantry Brigade. The headquarters and two squadrons of 36 Engineer Regiment accompanied them, the Commanding Officer, Lieutenant Colonel Geoff Field, becoming Force CRE. Other branches of the Corps were much involved. Survey quickly created new going maps to supplement the inadequate existing mapping of the islands. The Military Works Force became engaged in developing the facilities on Ascension Island, the key staging-post for the whole campaign. The Central Engineer Park and the Resources staff were fully extended arranging for the dispatch of the anticipated stores and equipment.

The landings, in the San Carlos area in Falkland Sound, were successfully accomplished but soon came under determined air attack in the course of which Staff Sergeant James Prescott was killed on bomb disposal duties (see page 182). 11 Field Squadron was credited with shooting down a Mirage with a light machine-gun. Among the most serious losses at sea was the SS *Atlantic Conveyor*, one of forty-five ships 'taken up from trade' for the war, going down with all the stores intended for the Harrier forward operating base and the Chinook helicopters in which the assault force was to be lifted forward for the attack on Stanley.

The San Carlos base and Harrier strip having been established (see illustration), operations proceeded that were to lead to the Argentinian surrender on 14 June. On the way to this successful outcome tough fighting had been experienced at Darwin and Goose Green on 28 May where a troop of 59 Squadron supported 2nd Battalion the Parachute Regiment. Another troop and one from 9 Parachute Squadron

marched north-east to Douglas settlement and Teal Cove, moving on foot with immensely heavy loads. The 5th Brigade with the reinforcements arrived at San Carlos on 2 June and moving forward through Bluff Cove made ready for the attack on Stanley. In preparation for this the main sapper task was identifying minefields and clearing safe lanes through them. The hard-fought attacks on the dominant ground to the west of Stanley, together with artillery and air strikes, settled the issue.

THE AFTERMATH

The euphoria of this triumph was tempered by the massive clearing up that faced the force, the need to secure the islands against any further attempt by the Argentinians and to restore public services and something approaching a normal life, albeit with a large military population. The immediate tasks, of which dealing with the proliferation of mines was one of the most serious, were undertaken by the squadrons of the task force while their relief could be put in place. This took the form of the specially established 37 Engineer Regiment, the headquarters and subunits to be provided in turn by regiments from UK and BAOR, and a CRE (Works) with detachments from 33 Engineer Regiment (EOD) and the RE Postal and Courier Services, all under a colonel CRE working to the two-star Commander Land Forces, British Forces Falkland Islands.

The main tasks were the operation of a military works area providing support for all three services, the upgrading of Stanley Airport, building a radar site on Mount Kent and West Falkland, the provision of eight hutted camps with

Below: The Sapper memorial, carrying the names of the nine members of the Corps killed in action.

Left: The Falkland Islands centrepiece, a Rockhopper penguin, a familiar sight on the islands, represented in silver by Stewart Griffith of London. The pedestal bears the Corps cap badge and the crest of the Falkland Islands.

Below: The South Atlantic Medal awarded to all those who took part in action on the Falklands and South Georgia. Those who only served on Ascension Island received the medal without the rosette.

the necessary services and the enhancement of the public utilities in Port Stanley to cope with the additional military population. In December 1982 during the South Atlantic summer, the CRE had under his command the CRE (Works), an

Above: The radar site on Mount Kent under construction in the period immediately after the war. (Painting by Tim Havers)

Above: Harrier Matting being laid at Stanley Airfield. In addition, the need for quantities of American AM2 contingency airfield matting had been foreseen early in the war as the only practical means of upgrading the runway to accept RAF fast jets such as Phantoms and Nimrods. A purchase was negotiated and was shipped via the UK ready for 50 Field Squadron (Construction) to begin work in mid-August. The runway was repaired, resurfaced and reopened for C130s by the end of the month. It was extended, complete with Rotary Hydraulic Arrester Gears, to accept the first Phantom on 18 October 1982.

Right: Coastels, the self-contained accommodation barges hired to house the huge influx of men that were needed for the post-war projects. The original vessel, *Safe Dominia*, was provided on hire from the Stena Line and was previously used by the oil industry. It provided for all the needs of 930 men. The best mooring that could be found, close to Stanley airfield, required a 1.7 kilometre approach road across peaty terrain, a berthing face capable of withstanding 300 tonnes and six massive bollards.

EOD squadron and 37 Regiment with six field and one field support squadron. The number of field squadrons reduced during the winter months, but the commitment continued until 14 March 1985 when 37 Engineer Regiment was disbanded and replaced by the Falkland Islands Field Squadron on a minimum manned strength and four-month *roulement* basis.

Above: Fox Bay Settlement after the war, showing the road and storage buildings.

44 | SEARCH FOR PEACE

THE LONG STRUGGLE

The smouldering sectarian hostility in Northern Ireland burst into flame in August 1969 when the annual Apprentice Boys' march in Londonderry was attacked by a Catholic crowd. A mass of civil rights grievances had been at issue for some time. The political ambitions of Sinn Fein and its military wing, the Irish Republican Army (IRA),[*] added fuel to the fire. Army strength in the province was increased in a situation that was well beyond the capacity of the Royal Ulster Constabulary to handle, but the hope was that the trouble would die down, so most of the early reaction was of a short-term nature.

From these small beginnings nearly thirty years of inter-communal strife were inflicted on the

[*]The term 'IRA' in this chapter includes both the Provisional IRA and the Official IRA, the militant splinter group formed in 1970.

province during which hideous acts of brutality were committed by both sides against each other and many innocent members of the public were caught in the crossfire. More than five hundred soldiers died in their attempts to bring peace, including twenty-nine Royal Engineers.

The intensity of the struggle to end the violence in the province ebbed and flowed with the political initiatives, the strategy of the republican and loyalist movements and the policies and actions of the security forces. Violence reached a peak in 1972 following the introduction of internment the previous year. 'Bloody Sunday' in January was followed by the imposition of direct rule in March, the breaking of a fragile ceasefire by the loyalists in May and Operation *Motorman* (see illustration) in July. The force level increased during the year from seventeen to twenty-seven

Left: Barricade and obstacle removal was a frequent task for sappers in Northern Ireland, represented here by a medium wheeled tractor with armoured protection clearing away a burning car in the Bogside, Londonderry, against the hostility of the local republican population. In a similar incident Corporal J Hamilton was wounded but succeeded in killing a gunman from the cab of his machine, for which he was awarded the Military Medal. (Painting by Terence Cuneo)

Right: Long Kesh internment camp, designed by 62 CRE (Construction) in 1971 to house 450 male internees in twenty-seven Nissen and twenty-three Twynham huts, with five permanent buildings. 48 Field Squadron built the camp between 29 June and 19 September 1971. The camp was later extended over the following winter for another 270 internees. It was renamed Her Majesty's Prison Maze in 1972 when the first convicted prisoners were taken in. The photograph was taken just before dawn.

Above: The 1962 General Service Medal with Northern Ireland clasp.

major units in the infantry role. In 1974 a brief flirtation with power-sharing provoked the Ulster Workers Council general strike. By the mid-1970s the efficiency of the Army acting in support of the Royal Ulster Constabulary had forced the terrorists to modify their tactics. A long struggle ensued in which small well-armed groups would attack key targets while their political wing fed the republican communities with propaganda. The Anglo-Irish Agreement in November 1985 led to further atrocities towards the end of the decade and in the early 1990s the majority of the murders were inflicted by loyalists. As time wore on it became clear to the terrorists that victory over the security forces was not a realistic prospect and that their brutality was becoming counter-productive. The political pressure for a solution had never wavered and in December 1993 the Downing Street Declaration led to cease-fires from both of the protagonists.

SAPPERS IN THE FRONT LINE

The Corps was involved both as infantrymen and engineers. 33 Field Squadron of 37 Engineer Regiment was, in 1970, the first squadron to be used in the infantry role. They later became the resident squadron in the province. The next year 21 Engineer Regiment from BAOR became the first whole regiment to be deployed as infantry. Throughout the 1970s and 80s this grew to a regular commitment for the BAOR regiments, who also maintained a field squadron in Northern Ireland on *roulement* in the engineer role. This, with another from the UK, and the resident squadron, provided for one for each brigade area, added to as necessary when the operational need arose.

From the start, search techniques began to assume great importance. The CRE became responsible for all search matters. A search cell was set up commanded by an infantry captain

Left: Operation *Motorman* was mounted in July 1972, a peak year of violence after the 'Bloody Sunday' operation in January. By mid-summer several parts of Belfast and Londonderry had become virtual 'no-go' areas cut off from the rest of the cities by strong barricades. To re-occupy these areas and establish control, one of the biggest operations by the British Army since Suez in 1956 was mounted. A troop of four Centurion AVREs from 26 Armoured Engineer Squadron in Germany was moved to England and shipped to Londonderry in HMS *Fearless*.

Above: The Borucki Observation sangar in Cross-maglen in process of dismantlement in July 2000. The post was named after Private James Borucki who was killed nearby by a bicycle bomb in 1976. In the five years before this photograph was taken forty-four army installations had been removed in the process of 'normalisation'.

and every unit was required to have a trained search adviser and two teams of six. The Corps took on the responsibility for training in all counter-terrorist search matters for the Army as a whole with its expertise centred on Chatham.

Specialised construction engineering became part of the front-line work of the Corps. Long Kesh and Magilligan internment camps, the latter a conversion of an existing training camp, were both constructed as emergency requirements, Long Kesh eventually being developed into the Maze prison. But perhaps the most demanding commitment was the development of protection for security force installations against the increasingly sophisticated weaponry acquired

by the IRA. In the early days protection was usually of a temporary and often improvised nature. As time went on the resources of the Military Works Force, helped by valuable advice from civil industry through the Engineer and Railway Staff Corps, were able to develop new designs using special materials for the quick erection of fencing, blast walls, rocket screens and bomb-proofing.

A special case for these techniques was that of the South Armagh infantry and police bases at Forkill and Cross-maglen. Normally such a project would be carried out by contractors for the Department of the Environment, but such was the threat to civilian contractors working on security force projects that none was prepared to tender and the Corps took it on. There was the added problem of access through hostile republican territory only possible by convoys absorbing enormous military manpower, so planning had to be accurate and allow for large on-site store areas.

These projects brought much credit to the Corps, confirmed the post-Falklands experience of the importance of the construction skills within units and established principles and practice that were to pay off in the challenges of the Balkans and the Gulf campaigns. Operation *Banner* formerly ended on 31 July 2007 but Royal Engineers continued stripping out military facilities for long after that: *first in – last out.*

Above: An arctic smock, a welcome issue to troops in the Northern Ireland winter.

Above: Irish Republican Army propaganda handkerchief.

Left: The Joint Base at Middletown, built in the late 1990s, following an Army-led re-appraisal of the design of these posts against the threat of the Provisional IRA's Mk 15 'barrack buster' mortar. Features visible in this photograph are the kinetic energy screen designed to stop or detonate incoming missiles, the sacrificial top storey of the main building and screen fencing with blast walls. The DCRE (Works) at Headquarters, as the Technical Authority for Hardened Structures, played a major role advising and assisting the contractors during the design process, subsequently supervising trials at the Royal Armament Research and Development Establishment and finally recommending staff acceptance of the contractors' designs.

45 | MAKING STRAIGHT IN THE DESERT

The two wars in Iraq against Saddam Hussein (February 1991 and March–May 2003) spanned the period of development of the Corps from its Cold War shape to something approaching the expeditionary force mode that was seen as the British armed forces' principal future operational task. The force was based on 1st Armoured Division in each case but was larger in 2003. However, whereas in 1991 integration of close support had to be contrived in-theatre to meet the need, by 2003 the sapper order of battle had been restructured to provide brigades with their own regiments.

Moreover, the need for the coordination of all joint force engineer effort was apparent from operational experience as far back as the Falklands as well as from the Balkans. By 2003 it was also established that some of the sapper specialisations such as Engineer Logistics, Geographical Support and Explosive Ordnance Disposal were properly regarded as part of the mainstream business of the whole force, rather than as appendages outside the normal chain of command and information.

On 1 August 1990, Saddam Hussein's army invaded Kuwait. The United States and Britain agreed joint action to liberate the country and before long twenty-seven more countries had joined the coalition. During the month RAF Tornados deployed to Dhahran and a sapper reconnaissance party went to Saudi Arabia followed at the end of September by 39 Engineer Regiment and a CRE (Works) once the decision had been taken to send 7th Armoured Brigade. Their task was to provide support for the Royal Air Force and to prepare for the reception and deployment of the Brigade, effectively establishing the Force Maintenance Area. By the first week in January 1991, 1st Armoured Division with two brigades was deployed from BAOR. Intensive training took place with many unfamiliar problems to be overcome including multiple obstacles of mines, burning oil, ditches and simply living and operating in desert conditions.

On 15 January the deadline set by the United

Left: The 1991 Gulf War lasted one hundred hours on the ground. It was preceded by twelve weeks of preparation and training in the theatre and followed by many months of clearing up and restoration of public utilities in Kuwait. In the picture a field troop of 21 Engineer Regiment is laying anti-tank mines while in the background can be seen a burning oil well. (Painting by David Rowlands)

Nations Security Council for the withdrawal of Iraqi troops had passed. Six weeks of intense air attack was then mounted on Iraq's command centres, military installations and infrastructure. The ground war was launched on 24 February, preceded by further air and artillery bombardment on the Iraqi front line and reserve positions. One hundred hours later the Iraqis surrendered and cease fire was called on 28 February.

In the immediate aftermath of the war the Corps contributed considerably to the rehabilitation of public utilities in Kuwait City and to the massive battle area clearance tasks. A field squadron (EOD) was deployed for the latter. In April what remained of the Iraqi army moved up in force to the Kurdish region in the north, in an apparent move to pre-empt a Kurdish revolt. Tens of thousands of Kurds fled to Turkey, instigating a serious humanitarian crisis. A United Nations task force was quickly assembled for which the Corps provided two field squadrons, an EOD troop, a specialist team and a resources section. The sapper team arrived in Turkey on 21 April and were tasked with providing support for the refugees in the mountain camps and for that part of responsibly securing the safe haven to which the Kurds would return. Within five weeks the team had restored the situation and was able to return.

Twelve years later 1st Armoured Division was back in Kuwait preparing to cross the border into Iraq, this time as part of the United States-led coalition whose aim was to occupy the country and replace the regime of Saddam Hussein with a democratic government acceptable to the West. The British contribution was part of the American 1st Marine Expeditionary Force, who with their V Corps were to make the dash north to Baghdad while the British secured Basrah. Three brigades each of different character made up the Division: 3rd Commando Brigade, 7th Armoured Brigade and 16th Airportable Brigade with their sapper support, respectively 59 Independent Commando Squadron with 131 Independent Commando Squadron (V), 32 Engineer Regiment and 23 Engineer Regiment. The Logistic Brigade (102nd) was supported by 36 Engineer Regiment and 28 Engineer Regiment with 23 Amphibious Engineer Squadron (equipped with M3 amphibious bridging) and 29 Armoured Engineer Squadron. The United Kingdom Air Contingent was supported by 39 Engineer Regiment with three field squadrons. Altogether fifteen specialist teams RE deployed to Iraq, of which three were with the Air Contingent.

Above: Desert Combat Smock.

The American planners were concerned to pre-empt any attempt by the Iraqis to set ablaze the oil wells in the Al Rumaylah oilfields on the Kuwaiti border, as had happened in 1991. Their capture was assigned to a US Marine regimental combat team, but they were accompanied by Royal Engineer oil and Explosive Ordnance Disposal specialists.

The advance on Basrah, led by 7th Armoured Brigade, was preceded by an attack by 3rd Commando Brigade on the Al Faw peninsula, at the junction of the Shatt el Arab and the Arabian Gulf on 21 March 2003. This was successfully

Above: Combat Engineer Tractor in the work-up to the Gulf War.

Left: Sand berms under construction in the preparatory phase before the Gulf War. (IWM GLF 880)

Above: The Gulf Medal, 1990–91.

Above: The Iraq Medal, 2003.

Right: 32 Armoured Engineer Regiment entering Kuwait in February 1991.

accomplished and the Division itself moved on in stages towards Basrah. There the problem was how to achieve the surrender of the occupying force without creating too much collateral damage. This was achieved by gradually dominating the surrounding country and closing in on the city and, on 6 April, a three-pronged attack was mounted. After some stiff resistance, success was finally achieved on 7 April. Baghdad fell on the 9th and by 14 April a general ceasefire was called.

Then began the long haul to restore the country to normality, to create conditions in which elections could take place, to train Iraqi security forces and help in the restoration of public utilities. All this had to take place against the highly active opposition of insurgents opposed to American intervention, mostly taking the form of savage suicide bomb attacks directed against both the occupying forces and the fledgling police and army training establishments and recruits. The Corps contribution to this was a close support regiment, a works group, a field support squadron and an Explosive Ordnance Disposal Troop.

While the Corps was working in Iraq, as well as in Bosnia, Northern Ireland and elsewhere round the world a further reorganisation of the Army as a whole, involving reductions to a level of about 102,000 by April 2008, was being planned. The Corps, however, was to

Above: 516 Specialist Team RE designed this 56-mile fuel pipeline to support the Gulf War. It was built under their supervision by a company from the United States Army in fourteen days. It became known as the Kinghan Pipeline, named after the Officer Commanding who was killed in a traffic accident. The Team's other tasks included deploying emergency fuel-handling equipment on the airfields and designing forty-two steel storage tanks to help meet the full demands for fuel and water.

Royal Monmouth-shire Royal Engineers (Militia) at work in Iraq. Nearly a third of the sapper manpower deployed in Iraq was provided by the Territorial Army.
Far left: A section from 100 Field Squadron repairing Rumalyha bridge, which had been badly damaged in the bombardment that preceded the ground offensive. Left: Aldershot bridge, a double truss, single-storey Bailey bridge built on twenty-eight pontoons over the Shatt-el-Arab.

increase its strength and capability, recognising the fact that engineering must be fundamental to all plans and affects all three Services. It also reflected on the impressive way the Corps had risen to the occasion, as so often in its history, culminating ultimately in closer integration with other arms and joint command structures. To meet its responsibilities it had succeeded in maintaining that mixture of both professional and military skills throughout its rank structure, continuing to adapt to meet new situations and through its special genius being on hand to produce solutions to the questions posed by commanders at all levels. Operation *Telic* ended on 30 April 2009 but a British presence continued in training and support missions.

Left: The 500-man Temporary Deployment Camp built at Basrah International Airport as part of the immediate post-war project to provide some 4,125 bed-spaces-worth of tented camps, fully air-conditioned with flushing lavatories, hot showers and catering and welfare facilities. The camps were built under contracts supervised by 62 Works Group between July and October 2004.

46 | JOINT ENDEAVOUR

Glasnost (openness) struck the Yugoslav federation in June 1991 when Croatia and Slovenia declared their independence of Belgrade after the first multi-party elections since the death of Tito in May 1980. Belgrade concentrated efforts to reunite the federation by military action against Croatian separatists, which reached a bloody deadlock. The United Nations attempted to mediate and sent a force, largely made up of French and Canadian troops but with a British field ambulance and a troop from 3 Field Squadron in support. Bosnia-Hercegovina also tried to break free but, unlike Croatia and Slovenia, could not win international recognition. The Bosnian Serbs, dedicated to the Serbian-led federation, built up their military strength and tried to crush the Muslim community into submission. Sarajevo, the heart of the Muslim community, was put under siege and mercilessly bombarded from the surrounding hills. A dire humanitarian situation in the city developed. The United Nations decided to send only sufficient force to protect the efforts of relief agencies trying to reach the city. The 22,500-strong United Nations Protection Force (UNPROFOR) was formed with the British contribution of a battalion group (1st Battalion the Cheshire Regiment) and 35 Engineer Regiment with 519 Specialist Team RE (see illustration).

Above right: 'Operation Grapple 1 (Route Triangle)' depicts 35 Engineer Regiment at work on one of the complex of roads that lay between the main base at Split and the British operational area. This regimental tour in Bosnia ran from November 1992 to May 1993.

Below right: The Mostar Bailey Bridge had been built to provide the only crossing over the River Neretva after the destruction by tank fire of the historic and symbolic Stari Most bridge. The Bailey itself was then subjected to severe tank fire and had to be replaced by a two-span extra widened Bailey, the most difficult part of the task being the removal of the old bridge. The painting by Terry McKivragan shows that bridge being prepared for demolition by 61 Field Support Squadron. Thousands of residents of the battered city had to be evacuated for the demolition. The new bridge was completed without trouble and opened on 12 September 1994.

BOSNIA-HERZEGOVINA

35 Regiment was replaced by a single field squadron but it soon became clear that the situation needed firmer action by the United Nations. The civil war erupted, the level of violence rising between the communities. The British contingent was increased to brigade level with a whole engineer regiment in support. The *roulement* programme, catering also for Northern Ireland and the Falklands, had to be carefully orchestrated

Right: An AVRE returning from a task through Mrkonjic Grad, December 1996.

Right: The mine-protected vehicle 'Mamba' which arrived in Bosnia in the autumn of 1996 to ease the problem of mine reconnaissance, seen here in use by 21 Engineer Regiment.

Right: Durrant bridge over the Vrvas river near Jajce, Bosnia, where a section of a road bridge had been blown during a Serb offensive. The bridge is a seventeen-bay triple single Mabey and Johnson spanning fifty metres and was built by 20 Field Squadron in August 1996. Like several other projects named after sapper VCs, this bridge commemorates Sergeant Tom Durrant (see page 151).

to ensure that the right mix of units was made available while spreading the load to avoid excessive diminution of tour intervals. 35 Engineer Regiment's turn came round again in 1995.

During all these years, still ostensibly in support of the humanitarian aid agencies, the sapper tasks had been the classic ones of maintaining mobility throughout the area of operations by bridging, maintaining and upgrading roads, and constructing, providing services for and maintaining camps. The indiscriminate use of mines by the warring factions led to major problems for the mobility of the forces and dreadful incidents involving innocent civilians. Mine awareness training and the dissemination of mine information became a major commitment.

In 1995 the Dayton Peace Accord was signed under which NATO was authorised to produce an implementation force (IFOR, Operation *Resolute* to the British) to ensure compliance with the ceasefire and supervise the separation of forces. This was based on the Allied Forces Central Europe Rapid Reaction Corps. The British

Left: An observation tower under construction in a Serb enclave near Pristina in July 2000. Several of these were built to allow monitoring of Serb activity.

Above: The United Nations Protection Force medal for Bosnia.

Below: The NATO Medal for Macedonia.

element was essentially a division: 3rd Division was followed by 1st Armoured Division, each with two engineer regiments of two or three squadrons and a strong element of specialist support. Several non-NATO countries also contributed to IFOR including Hungary and Romania. There was a prodigious amount of engineer work to be done in the aftermath of the war. As an emergency many temporary pontoon and assault bridges had be laid to provide immediate communications. The overall IFOR plan led to the eventual replacement of some sixty bridges.

While the NATO intervention had achieved its aim, the situation after IFOR was still delicate. The United Nations agreed on a 'stabilising force' (SFOR, Operation *Lodestar* to the British), half the size of IFOR, with which to coax the communities into normal life and prevent further hostilities. A single engineer regiment met the engineer commitment.

KOSOVO

Frustrated in Bosnia, the Serbs began to harass the Muslim (ethnically Albanian) population of Kosovo, one of the two provinces of Serbia itself. Outrages became commonplace and desperate columns of refugees started to make their way to Albania. It became increasingly clear that military action against Serbia would become necessary if a massacre were to be avoided. A verification mission from the Organisation for Security in Central Europe was sent in to observe the facts and found no evidence other than that the Albanian population was in peril.* NATO now

*The Chief of Operations of the Mission was the retired sapper Major General John Drewienkiewicz.

Right: Bosnia and Herzegovina map of the Ceasefire Lines and IFOR deployment under the Dayton Agreement, March 1996.

prepared to bombard Serbia from the air to force compliance with the United Nations demands followed by a ground invasion to establish a peace.

The bombing started on 24 March 1999 and KFOR (Kosovo Force) entered the province on 12 June. The force was under the overall command of Lieutenant General Sir Mike Jackson, commander of the Allied Forces Central Europe Rapid Reaction Corps. After the initial entry, fortunately unopposed, the British component settled down to brigade size, the first sapper field unit being 69 Gurkha Field Squadron. Soon it became apparent that the major task lay with restoring public utilities, particularly the electricity system, and with setting up camps under the newly devised expeditionary force scheme for temporary field accommodation. In all, thirteen camps were built by contractors working to the CRE's Works Project Management Team.

By 2004 when the Balkans commitment came to an end the Corps could look with pride at the

MINE HAZARD

Operations in the Balkans coincided with the heightening of world-wide concern about the proliferation of mines particularly in countries where insurgencies had been in progress for long periods of time, especially in Afghanistan, Cambodia and many parts of Africa. Under pressure from public figures, including Diana Princess of Wales, the Ottawa Convention came into force on 1 March 1999 aimed at reducing the casualties among innocent civilians in such countries. By then all British stocks of anti-personnel mines had been destroyed.

In September 1997, the Mine Information Training Centre was set up under the Battlefield Engineering Wing at Minley for training for all three services in mine awareness and a Grade 1 staff appointment for Humanitarian Mine Action was added to the staff of the Director of Military Operations.

Right: Mine Information and Training Centre logo.

contribution they had made to bringing some order to what had seemed a decade earlier an almost impossible situation. The pressures on the units of the Corps had been severe; but the experience gained and the contribution of that experience to the evolution of the sapper order of battle and further integration with all arms and the other two Services was to pay off in the future.

Above: The NATO Kosovo Medal.

Above: Entry into Kosovo was painted to commemorate the Corps' participation in the entire Balkans operations. It depicts the sappers leading the ground forces through the Kaçanik defile into Kosovo (see text). The two poppies in the left foreground were included in memory of Lieutenant Gareth Evans and Staff Sergeant Balaram Rai, both of the Queen's Gurkha Engineers, who lost their lives in follow-up operations.

47 | ON A DISTANT PLAIN

The Al-Qaeda attacks of 11 September 2001 on New York and Washington DC shocked the world and led to military intervention in Afghanistan, the centre of Al-Qaeda training infrastructure. Within two months, the United Kingdom joined the United States in its response. The British deployment to Afghanistan and to fight international terrorism began as Operation *Veritas*. Operations such as *Fingal*, the UN-sponsored deployment to provide security and stabilisation in the Kabul area, and Operation *Jacana*, the Royal Marine Commando deployment to support the fight against the Taliban and Al-Qaeda in Khost province on the border with Pakistan, fell under the Operation *Veritas* umbrella. In December 2001, under United Nations Security Council Resolution 1386, the International Security Assistance Force (ISAF), led by NATO, was established with the aim of supporting the Afghan administration, providing security and

training the Afghan security forces. From the outset, British forces were at the forefront of the international response to the 9/11 attacks.

Operation *Fingal* employed elements of 16 Air Assault Brigade, including 9 Parachute Engineer Squadron, which had only recently returned from Macedonia, as part of 36 Engineer Regiment Group. The 2 Parachute Battalion battle group was responsible for securing the southern part of Kabul and Sappers undertook essential enabling work that provided infrastructure and force protection. Whilst a substantial part of the mission involved security tasks, the battle group trained an Afghan National Army battalion of 600 men and hosted a football match between ISAF and Kabul United. For Operation *Jacana*, Royal Marine Commandos used Bagram Airbase as their base of operations. Here, Sappers cleared a minefield and established accommodation to allow operations to get underway. This work allowed the task force to conduct mopping-up operations throughout the mountain valleys bordering Pakistan. At the end of 2002, Operation *Veritas* closed and all UK operations in Afghanistan fell under the umbrella of Operation *Herrick*.

In its early stages, the campaign in Afghanistan was very different from its later development. The situation remained relatively stable for the initial stage of the NATO mission. Sappers were deployed to Kandahar and Mazar-e-Sharif to establish an airfield and to carry out missions under the Provincial Reconstruction Teams

Above: Logistic Support Bridging. Sappers in all wars have turned their hands to bridging both in the assault and in support of logistic lines of communication. Afghanistan was no exception.

Below: ABLE General Support Bridging. This provided a temporary crossing in support of immediate operations and was replaced by MALVERN Bridge. (see page 233)

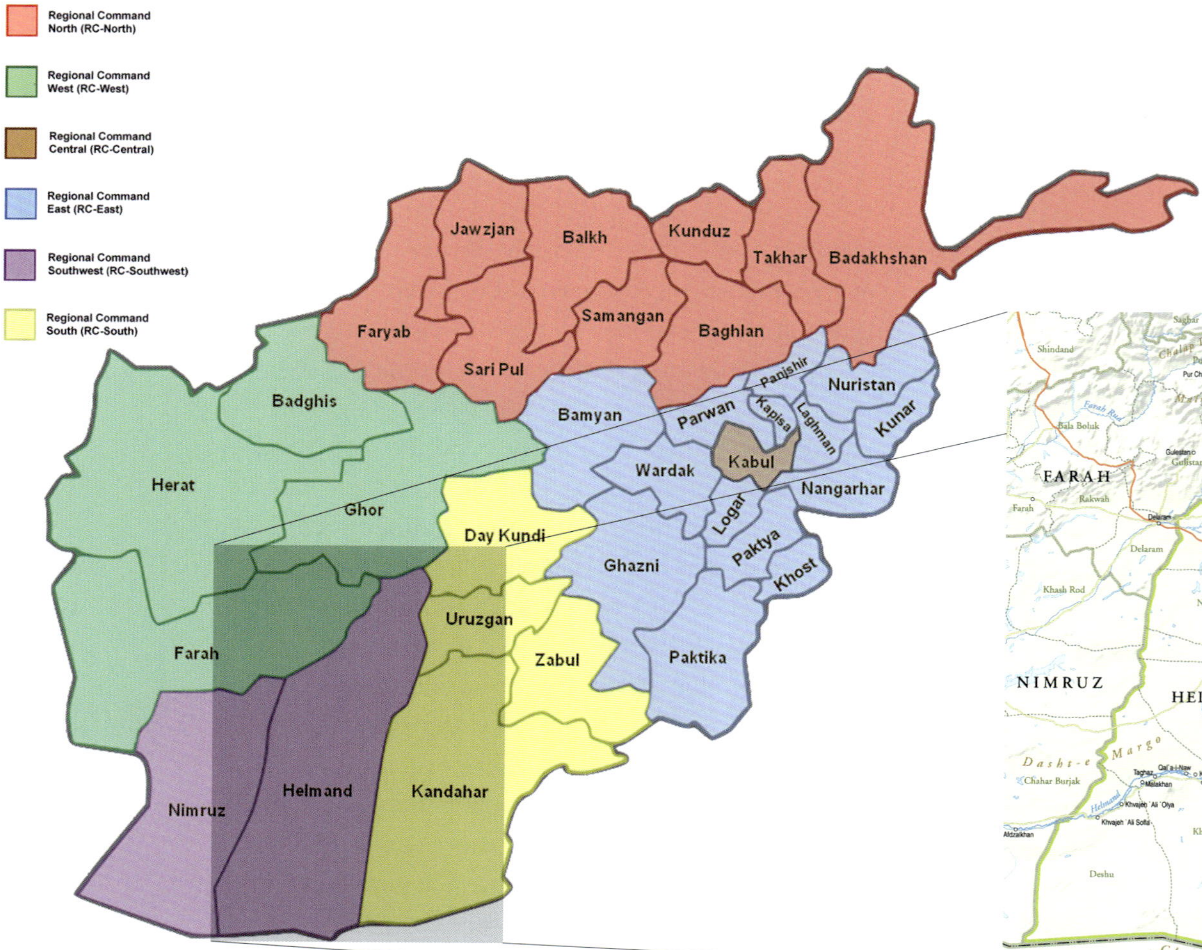

Regional Command
North (RC-North)

Regional Command
West (RC-West)

Regional Command
Central (RC-Central)

Regional Command
East (RC-East)

Regional Command
Southwest (RC-Southwest)

Regional Command
South (RC-South)

Right: Map of Afghanistan, showing the regional command structure and the major provinces.

Below: Map of Helmand Province

Prepared by David Kirby - US State Department, Public Domain

(PRT) Programme. Alongside the operations in Kabul, sweeps aimed at clearing out Al-Qaeda and Taliban forces continued. Yet, dissatisfaction was bubbling away under the surface of Afghan daily life and over the border in Pakistan. This simmering tension soon boiled over into a full blown insurgency as NATO forces extended out from the Kabul area.

After another UN resolution in 2004, an expansion of the ISAF mission was approved. The enlarged NATO campaign was designed to roll out in four stages. Each stage brought the mission into another region of Afghanistan. Stage one moved North in 2004 and then the step-up continued anti-clockwise until stage four when ISAF took responsibility for NATO and Coalition forces throughout Afghanistan. In 2006, British forces deployed into Helmand province as part of the expansion of ISAF into the south of Afghanistan and, for the next eight years, ISAF fought against a fierce Taliban insurgency. The counter-insurgency campaign in Helmand was based on two thrusts of action. The first was to find and fight the

enemy, to deny the insurgents control of territory and of the population. The second, but just as vital, was to improve the lives of the civilian population in order to undermine the influence of the Taliban. The fighting against the Taliban in Helmand was of an intensity not seen in many years and Sappers made essential contributions from first to last.

Below: Sappers in heavy contact at a helicopter landing site.

Above: Camp Bastion: Camp Bastion was a major construction project that lasted a number of years as the operation expanded. It grew to include a runway capable of taking wide-bode jets and a large military hospital.

CAMP BASTION

Planning for the expansion of the NATO campaign into Southern Afghanistan began in 2004. A major part of this effort was providing somewhere for British and allied forces to operate from. This led to the construction of Camp Bastion in a remote part of the Afghan desert. The project was led by 62 Works Group with 39 Engineer Regiment. In 2006, Engineers turned the temporary tents into a more permanent camp. Whilst the project was led by the military, it involved significant contributions from civilian contractors and the Corps was honoured by a British Construction Industry Award for their efforts in constructing the camp. By 2007, the runway that led to Camp Bastion being the fifth busiest UK-run airport in the world was in place. As the campaign in Helmand wore on, the camp expanded to house American, Danish, Afghan, and Estonian troops. The hospital in Bastion was built by Sappers from 170 (Infrastructure) Engineer Group. In the best traditions of the Royal Engineers, Sappers were on the ground laying the foundations of Bastion and were the last to leave as they supervised the drawdown. Returning the substantial amounts of material and supplies was a task as complicated as putting them there in the first place. The operation to dismantle Bastion in 2014 saw the decommissioning of a camp that covered some 50 square kilometres.

Below: The Afghanistan Medal from 2001

Outside Camp Bastion, Sappers undertook a variety of force protection projects. A key element of counter-insurgency campaigns is known as the spreading of 'ink-spots'; friendly forces are moved into a location and they secure it. From there, they extend their control of the area and deny it to the enemy. Once security is established, development can take place to win the hearts and minds of the local population. In order to support this, Royal Engineers were deployed to build forward operating and patrol bases used by ISAF and Afghan forces.

COUNTER IED

As the campaign in Afghanistan progressed, the methods of the Taliban insurgents changed. At the start of the deployment to Helmand, the Taliban attempted conventional attacks on coalition forces. This resulted in pitched battles in which the insurgents usually suffered significant losses. In a change of tactics, they began to use improvised explosive devicwes (IEDs) on an increasingly large scale. The effects on ISAF forces were profound. IED strikes became the primary source of casualties. Brigadier Shaun Burley's 2008 review found that 79% of British casualties were from IEDs. This situation needed a response and Sappers were destined to be an integral part of the fight against the IED.

The counter-IED campaign was focused on three principal aspects: attacking the network, defeating the device, and preparing the force. Each one of these elements was key to countering the IED threat. In order to attack the network, C-IED specialists would attempt to disrupt an IED before it detonated. This would provide them with intelligence regarding the construction of the device. Other elements were location and vehicle searches. In essence, attacking the network was about gathering intelligence on the devices and the people who were making them. With this information Sappers and all others involved in fighting IEDs were better equipped for the task at hand. Defeating the device covered all aspects of finding and then disrupting them before they had the chance to do any harm. Detection involved everything from metal detecting equipment to military working dogs. The Talisman system was an essential part of this and is discussed in the sidebar (below). Preparing the force was all about training ISAF and Afghan forces in how to detect IEDs. Whilst the campaign pressed on, Afghan-style villages were constructed in the UK and Germany for immersive training. As part of wider changes to the Army brought in to fight the campaign in Afghanistan, an Engineer regiment was re-roled to carry out the search function. Geographic (Geo) elements of the Corps supported C-IED and other ISAF operations, and continued to expanded their vital role within the

Corps and the wider Army. With the ability to fight back against IEDs and to manoeuvre more freely, British forces in Helmand could continue their battle against the Taliban.

INFRASTRUCTURE

Combating the Taliban was only one part of the campaign. Alongside fighting the insurgents, Sappers undertook a huge variety of projects to support the Afghan population. These tasks varied in scale but were all carried out with the

TALISMAN – THE UK ROUTE PROVING AND CLEARANCE CAPABILITY

TALISMAN was the British Army's route proving and clearance system. It evolved from in-theatre responses to IED strikes on vehicles. Once developed, it became an effective way of countering the threat of IEDs. Three different types of specially adapted vehicles, Mastiff, Buffalo, and the High Mobility Engineer Excavator (HMEE), were used alongside TALON drones and DRAGONRUNNER robots. Of course, the key elements were the personnel who had undergone training on the relevant systems. Each *Herrick* deployment came with a TALISMAN squadron that provided vital support to the mobility of the larger battle group. The lead vehicle was a Mastiff equipped with a mine-roller so that it was able to survive most IED strikes without being destroyed. Other vehicles in the column were kitted out with devices that disrupted signals that insurgents used to detonate IEDs remotely. The deployment of TALISMAN squadrons reduced ISAF and Afghan casualties on operations whilst providing an asset that enabled manoeuvre and mobility. The Sapper teams that dismounted from vehicles to disarm IEDs were codenamed BRIMSTONE. Their expertise and courage in taking on the IED scourge was vital to the successful conduct of operations in Afghanistan.

Above: The last TALISMAN vehicle shows the way for the CLP.

Below: A TALISMAN troop, ready to deploy.

Below right: A search team member demonstrating persistent courage at the sharp end.

Below: MALVERN Bridge
Improving the lives of the civilian population is a key part of a counter-insurgency campaign. Engineers built different sized bridges throughout Helmand in order to improve the lives of the Afghan people. The largest example was the Malvern bridge, built in 2012 across the Nahr-el-Bughra Canal. The specification was for a 67m MLC 40 bridge with level ramps so that the bridge could also be used by civilian traffic. The bridge was a bespoke Mabey LSB Compact 200.

objective of improving the lives of the Afghan people. A significant example was the operation to deliver a turbine to the Kajaki Dam hydroelectric power station in 2008 to increase its generation capacity. Operation *Oqab Tsuka* (*Eagle's Summit*) involved planning, logistics, and route proving and clearing. The convoy stretched out over two and a half miles and wound its way over uneven terrain and roads within the Taliban area of operations, hostile territory in every sense, before delivering the turbine. The Sapper effort was led by 23 Engineer Regiment (Air Assault), supported by the rest of the Engineer Task Force.

Alongside such headline operations, a wide range of infrastructure work was carried out for the benefit of the Afghan people. These tasks varied from the construction of police stations, schools, hospitals to all kinds of bridges. Bridges provided not only mobility for the forces fighting the Taliban but also made the lives of Afghan civilians easier. The Malvern Bridge in Helmand province spanned 67 metres and was the largest single span bridge constructed by the Royal Engineers since the Second World War. Most of these tasks were part of the PRT scheme and by units from 170 (Infrastructure) Engineer Group. A vital part

of these tasks was the use of local contractors which provided a good opportunity to boost the relationship between the military and the civilian population. Contractors were also employed to build roads, such as Route Trident, under the guidance of the PRTs.

END OF COMBAT OPERATIONS

Although the campaign in Afghanistan was long and arduous, Sappers performed their duties with characteristic dedication. Some were rewarded for their bravery and others made the ultimate sacrifice. The ferocious insurgency faced by British and other NATO soldiers in Helmand exceeded some expectations. Much through the efforts of Sappers across many different elements of the campaign, British forces were able to be withdrawn, having handed over security to the Afghan National Army and Police. Many members of the new Afghan army had been trained by Royal Engineers. During the period of operations in Afghanistan, successive British governments had reduced the size of the Armed Forces. Enhancements made to the Army Reserve, however, showed how vital volunteer elements had been in Afghanistan. As Operation *Herrick* ended and Operation *Toral* began, the United

Kingdom's contribution to NATO's enduring Afghan training mission, both Afghanistan and the Royal Engineers were very much changed since operations began thirteen years before. In subsequent years, the fight against international terrorism continued elsewhere on the globe. On operations such as Operation *Shader*, the air support mission against Islamic State targets in Iraq and Syria, Sappers continued to employ their training and skills.

Below: New equipment
Many new pieces of equipment were first deployed in Afghanistan. Some were bought in response to the demands of the campaign, Urgent Operational Requirements (UORs); others were adapted from existing stock. The Trojan AVRE (below left) had its first deployment on Operation *Moshtarak* after being flown in from the United Kingdom. TALON (below right) was one of a number of robots bought to fight the IED threat.

Right: Sapper skills
The campaign in Afghanistan called on all key sapper skills: Soldier, Combat Engineer and Tradesman. The images show these talents in action: clockwise – assault breaching in close support of the infantry; a search find; a super sangar at the entrance to a forward operation base; and the building of one of many culverts throughout the operation.

48 | A HUMANITARIAN MISSION

The expertise of Royal Engineers ensures that they are able to respond to a great many challenges and can be quickly on hand to provide humanitarian assistance, often at the request of the Foreign Office in support of defence relations. They are called upon, therefore, in times of need at home and abroad. The humanitarian missions undertaken by the Corps have varied from bridge building in response to flooding, rebuilding harbours devastated by storms and building camps following earthquakes. These operations have come in all shapes and sizes and required different scales of effort and manpower. When the main defence effort was focused on the campaign in Afghanistan, the scope for other operations waned. As the drawdown in Afghanistan neared its conclusion, more elements of the Corps were available to be deployed on Humanitarian Assistance and Disaster Relief (HADR) missions. Although Sapper involvement in these operations never completely ceased, it was reduced during the periods of heaviest commitment to Afghanistan as resources were stretched. The increase in operations of this type was a return to previous norm, rather than a new development in the history of the Corps.

At home, the Corps responded to the effects of increasingly extreme weather conditions in the UK with support to flood-hit communities throughout the country. Over a number of years, the Corps' response to flooding was tested.

Left: The Cavendish Bridge at Shardlow. The Corps' long tradition of assistance to the civil community in flood relief is marked by a large number of temporary bridges, repairs to dams and other structures. In this case, during the devastating floods in the North of England in 1947, a 100ft, Class 70, triple-triple Bailey Bridge was constructed by the Officer Cadet Training Unit (OCTU) in Newark for the Ministry of Transport, with a life expectancy of 5 years before a permanent replacement could be built.

FLOOD RELIEF

Cumbria Floods 2009. Following the storm that occurred during the night of 19-20 November 2009, which produced 316mm of rain in just 24 hours and damaged all 3 bridges across the River Derwent in Workington, the Royal Engineers were called on to replace the North Side Bridge. The team constructed a 52m span footbridge across the River Derwent, converting a green-field site into a working crossing in just 13 days. The bridge was named the Barker Bridge after PC Bill Barker who lost his life there on 20 November.

Below: After heavy rainfall in December 2013 and twice the average rainfall in January 2014, the south of England faced considerable flooding. The Army and the Royal Engineers assisted in many areas. In the town of Romsey in Hampshire, the River Test burst its banks and the Fishlake Stream running directly into the centre of Romsey on a raised earth embankment was overtopping. There was already extensive flooding of roads and businesses. However, the major concern was further damage to the earth embankments which were perilously close to collapse; failure would have been catastrophic, causing uncontrolled flooding throughout the centre of town. The Environment Agency requested military assistance and 600 military personnel deployed to the town as part of Operation *Pitchpole*. Whilst the rest of the military laid thousands of sandbags, the Royal Engineers were set the challenge of constructing two weirs in the flooded rivers to reduce the rate of water entering the Fishlake Stream. The first was made of stripped down Hesco Bastion acting as gabions and the second an improvised weir using scaffold braced to an existing footbridge.

Below: Fishlake Stream Restriction Scheme, Romsey

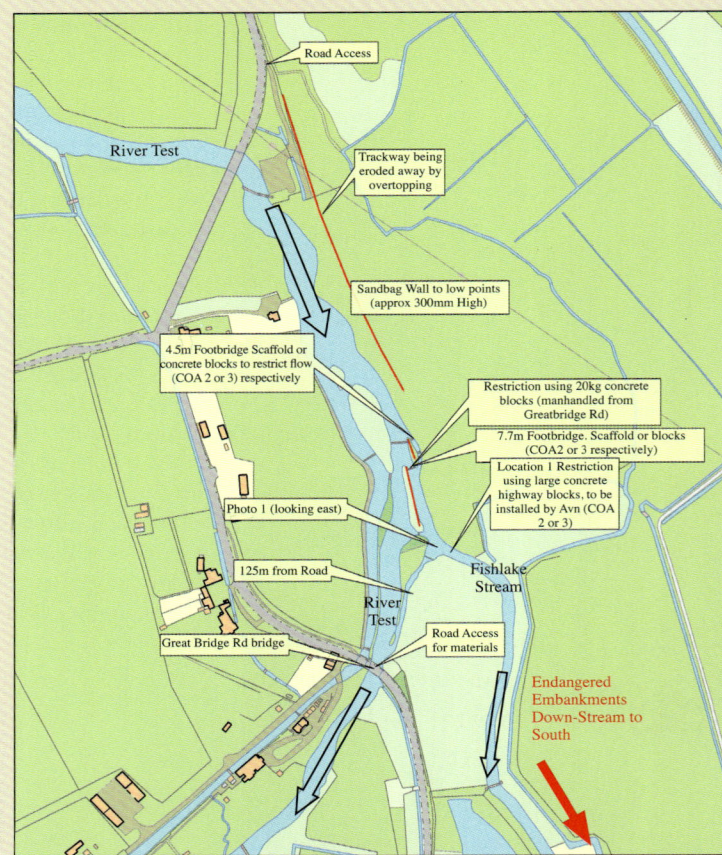

Above: The collapsed Northside Bridge at Workington on 20 November 2009.
Below: The completed Barker Bridge.

Right: Sappers building the gabion dam.

Right: Completed scaffold structure acting as a Weir; a 300mm local drop in water level is visible.

Above: One of 3 sites in the Falkland Islands that formed Project *Anemoi*.

Right: The construction of a taxiway at RAF Lossiemouth, as part of Op *Adana*, to allow RAF Typhoons to meet their readiness targets.

In the summer of 2007, Gloucestershire saw unprecedented rainfall that led to widespread flooding. Tewkesbury Abbey was surrounded by water whilst flooding threatened an electricity substation at Waltham. Intervention by Sappers ensured that the floodwaters did not enter the facility preventing a more serious incident. In 2009, floods washed away and damaged bridges in Workington, Cumbria. Sappers constructed a 52 metre footbridge to reunite the community whilst the other bridges were repaired. Sappers were again on hand to provide support to the emergency services as they dealt with flooding

at Romsey in Hampshire in 2014. Facing rapid water flow rates, the ad hoc team, consisting of engineers from 170 Infrastructure Support Group and 22 Engineer Regiment, were tasked with constructing a weir to divert floodwaters away from the town. With improvisation and ingenuity typical of the traditions of the Corps, the team achieved its mission. This was an interesting example of cooperation between the military and civilian authorities. It was noted at the time that the presence of Fire Service divers enabled the Sappers to focus on the task of constructing the weir as they were safe in the knowledge that the divers were on hand. In 2015, 21 Engineer Regiment inspected over one hundred bridges in Cumbria following Storm *Abigail*. The ability of the Royal Engineers to deploy to floods and to provide essential support to civilian agencies has been and will continue to be tested by storms in the United Kingdom.

Major construction tasks, which once had been a regular activity for the Royal Engineers, had become far less frequent. This situation changed following the drawdown in Afghanistan. In 2014, the Corps responded to a tender from the Defence Infrastructure Organisation to renew remote accommodation in the Falkland Islands. This became Project *Anemoi*. The objective of *Anemoi* was to build new accommodation at radar stations. With the lengthy logistics chain and austere conditions, it was a good opportunity for the Corps to practise project management and construction skills on a scale that had not been used for several years. In unreliable weather and with a six-month lead-time, the Corps began to construct the accommodation blocks. From the outset, there were challenges in getting material to the sites and working in extreme weather conditions. The task was completed over several *roulements* and provided excellent trade and project management experience.

53 Field Squadron was tasked in the summer of 2015 to build a new 250m x 16m wide taxiway at RAF Lossiemouth to enable the Quick Reaction Alert (North) Typhoons to meet their mandated readiness times. This complex construction project, known as Op *Adana*, tested the technical ability and endurance of plant operators, technical tradesmen and combat engineers, all

Right: Op *Morlop* involved the clean-up of a number of contaminated sites around Salisbury, including dismantling the entire roof of a house whilst wearing protective clothing.

Above: Medal for the Ebola crisis, inscribed on the reverse *For Service West Africa Ebola Epidemic.*

Below: A Sapper from 59 Independent Squadron Royal Engineers at work building shelters in the Bagh District of Kashmir, as part of the UK's aid to Pakistan after the earthquake on 8 October 2005.
Image by POA Phot Ian Richards; ©Crown Copyright 2020

working long days to ensure the highest standards were met and that the task was completed in 13 weeks, ahead of the Ministerial deadline. The task was a huge success and saved the defence budget nearly one million pounds.

In March 2018, an attempt was made on the lives of Sergei and Yulia Skripal in Salisbury. Those responsible for the attempt used a potent nerve agent known as *Novichok* (the Russian word for newcomer), manufactured in the Russian Federation. Three months later, in June 2018, Charlie Rowley and Dawn Sturgess were hospitalised after showing symptoms of nerve agent poisoning. Dawn Sturgess died eight days after admission. The threat posed by contamination from this persistent nerve agent was very serious. In response to the initial incident, Sappers from 22 Engineer Regiment deployed to aid the clean-up operation. This was known as Operation *Morlop*. After a short recce, the complex work began. Although the task was finished ahead of schedule, the operation still lasted 337 days with significant time spent in protective equipment during both winter and summer months. The difficult working conditions added to what was already a sensitive task; there were multiple sites of poten-

tial contamination, which had to be dealt with. The largest single task involved the dismantling of an entire roof section of a house. In addition, the full spectrum of Sapper skills was required to decontaminate items varying from furniture to water pipes. Through the efforts of the Royal Engineers, life in Salisbury was able to return to normal after a shocking event with long lasting repercussions.

Disaster relief operations are not limited to the UK. Sappers have been deployed throughout the globe to provide support to people in dire situations. Following earthquakes in Nepal and Pakistan, Sappers deployed to provide relief to the local population. 59 Independent Commando Squadron and elements of 42 Commando Royal Marines deployed to aid the international response to the 2005 Kashmir earthquake. This event caused upwards of 80,000 deaths and led to the displacement of 2.8 million people. After deploying to Bagh on Operation *Maturin* in November 2005, Sappers built health centres and schools and provided many other forms of help, where needed. Mountain training proved useful as the Commandos were operating at altitudes between 4,000 and 7,000 feet. Major Nigel Cribb, OC of the operation, was personally thanked by the Pakistani Prime Minister at a reception at 10 Downing Street in March 2006. In 2015, a 7.8 Richter scale earthquake hit Nepal leading to 8,000 deaths and substantial destruction. 36 Engineer Regiment, with its two Gurkha Engineer Squadrons, was deployed to aid in disaster relief and reconstruction efforts.

Below: Gurkha Sappers from 36 Engineer Regiment on Op *Layland*, working through the night, building temporary shelters in Nepal after the 2014 earthquake.

On Operations *Layland* and *Marmat*, Gurkha Sappers were able to help their home country through building schools, health centres, and accommodation for retired Gurkhas.

As the Ebola virus spread through West Africa in 2014, Sappers were deployed to Sierra Leone on Operation *Gritrock* to supervise construction of specially designed treatment centres. These centres were built by local contractors to designs and specifications put in place by the Sappers on the ground. Given the highly infectious nature of the virus, it was vital that every effort was made to limit transmission of the disease.

Weather conditions were the foe in operations in the Philippines and the Atlantic island of Tristan da Cunha. Following Typhoon *Haiyan*, which struck the Philippines on 7 November 2013, HMS *Daring* was diverted to provide disaster relief. Sappers from 24 Commando Engineer Regiment and 42 (Geographic) Engineer Regiment led the Sapper contribution to the UK's disaster response. Previously, a smaller but equally vital operation had taken place on the small Atlantic island of Tristan da Cunha, a British Overseas Territory. In 2008, Sappers were sent to the island to conduct repairs to Calshott Harbour in Edinburgh of the Seven Seas, the island's main settlement. Although the reinforced harbour has since been battered by Atlantic storms, it has withstood the conditions. Securing the island's lifeline has demonstrated the success of Operation *Zest*. These two operations, vastly different in scale and scope, show how the humanitarian efforts of the Corps support people all over the world.

The end of major operations in Afghanistan led to increased capacity for UK Forces to return to supporting other UN, multi-national and humanitarian missions; this became a key tenet of the 2015 Defence Review. Given their abilities in enabling and force protection work, Sappers were well placed to lead these deployments, such as Operations *Catan*, *Branta* and *Trenton*. On these operations, Sappers were required to demonstrate a variety of expertise and to maintain an adaptable posture in facing dynamic conditions. Operation *Catan* was the British commitment to the UN mission in Somalia; here, Sappers carried out force

protection tasks. Sappers took on a similar role when deployed to the Sinai in support of the Multinational Force and Observers on Operation *Branta*, the enduring mission there.

Operation *Trenton* was a longer-lasting operation in the world's newest nation, South Sudan. The UN Security Council identified the instability in South Sudan as a potential threat to international peace and security and voted to establish a UN mission there. South Sudan was mired in a civil war and faced numerous problems such as inter-ethnic tensions, large numbers of displaced persons and the challenges of building a successful state following its independence in 2011. Over the course of 2017 to 2019, regiments from the Corps deployed in support of the United Nations Mission in South Sudan (UNMISS) as the 300-strong UK Engineer Task Force. Operating from UN bases in Malakal, Bentiu and Juba. Sappers built two hospitals, upgraded roads, schools, prisons and bridges. Given the volatile nature of the political situation and previous attacks on UN Peacekeepers, force protection works were also essential. Through the construction and upgrades to over 15km of roads, Engineers helped prevent sexual violence against women collecting food and water. Operation *Trenton* was the largest

Left: One of the treatment centres designed and project managed by Royal Engineers in Sierra Leone on Op *Gritrock*, during the Ebola virus outbreak in 2014.
Image by Lieutenant Steve Dunning RN; ©Crown Copyright 2020

Above: Medal for Op *Branta*, inscribed on the reverse *United in Service for Peace.*

Left: UK troops were deployed on Op *Catan* in Somalia in 2016. The troops were part of the UN mission in East Africa to help bring stability to the region. The force's mission included force protection and the provision of Engineering support. This was the largest deployment by UK troops on a UN mission since Bosnia.
Image by Cpl Pete Brown; ©Crown Copyright 2020

UK deployment of peacekeepers in two decades and further demonstrated the ways in which the Corps contributed military aid abroad. Undoubtedly, the contribution from UNMISS and the UK Task Force prevented further casualties in a region that had seen much death and destruction. By ensuring that aid could be delivered quickly and safely, the UN force averted further humanitarian catastrophes.

Military Aid to the Civil Powers (MACP) and Community (MACC) both continue and, together with HADR, will always be part of the significant contribution the Corps of Royal Engineers makes in times of crisis. The hard-gained experience of these missions will, therefore, continue to serve the Corps into the future. The need for military aid in response to humanitarian disasters and incidents will not abate in years to come. What is clear from these operations is that the Corps retains the expertise and flexibility to provide military aid at home and abroad in a variety of different situations – *Ubique*.

Operation *Trenton*. Royal Engineers deployed to South Sudan as part of the UN peacekeeping mission, providing engineering support to the Force. Force protection was a key theme, with the construction of sangars and solar-powered security lighting, and the provision of buildings and other facilities, including a pipe range at Malakal.

**OPERATION
*TRENTON***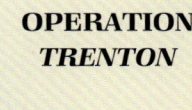

VII
FAMILY AFFAIRS

49 | ESPRIT DE CORPS

Today's Corps has inherited a strong tradition, safeguarded in the institutions that make up the 'Corps Family'. Its allegiance is to the Sovereign as Colonel-in-Chief, represented on formal occasions by the Chief Royal Engineer.

The Corps Family encompasses all those who have worn the RE Cap Badge and their families. It is strong and flourishing, supported by focused, modernized institutions that have evolved to meet the changing needs of the Corps, which includes all those organisations such as messes, societies, clubs for sports and other activities to which members of the Corps, serving and retired, regular and reserve, belong. Through these, the Corps publications and web-site and, above all, through service together in peace and war, is built up that *esprit de corps* that survives long after the end of active service; hence the saying, 'Once a sapper, always a sapper'.

CONTROL AND FINANCE

The Institution of Royal Engineers, the Royal Engineers Association, the Royal Engineers Central Charitable Trust and the Royal Engineers Headquarter Mess are all separate charities, governed by their trustees, but reporting and coordinating their activities through the Chief Royal's Board.

Corps family affairs are financed by the private subscriptions of all ranks of the Corps who also subscribe separately to the various organisations to which they belong. Until 1987 the funds of the different elements of the Corps were managed separately. Since then all funds have been centralised under the Corps Treasurer; benevolence, as disbursed by the Royal Engineers Association, always having first call on the funds.

Corps governance comprises the RE Affairs Committee, the Council of the Institution of Royal Engineers and the Trustee Board of the Royal Engineers Association, each chaired by a Colonel Commandant. The latter two bodies are separate charities and are governed, respectively, by a Royal Charter and a Deed of Trust.

The Royal Engineers Central Charitable Trust receives income for all Corps activities that meet the terms of the Trust Deed to support Corps benevolence, sports and games, *Sapper Magazine*, a band, a museum and *esprit de corps* grants.

Until August 2002, the Corps' investments were managed by the Corps Investment Committee, which reported to the Corps Finance Committee and each of the Corps' charities. Although the accounts flourished, the need for professional

Below left: The Ravelin Building, the home of the Royal Engineers Museum with which the Royal Engineers Library is to be located, was built as the Electrical School in 1905. At that time the Corps needed large numbers of electricians, a considerable proportion for the submarine mining service and searchlights. The towers were designed for training on searchlights in their coastal defence role. After the Corps lost the responsibility for submarine mining the building was used for a wide variety of purposes until the Museum and the offices of Regimental Headquarters Royal Engineers moved there in 1985. The open courtyard was covered in 1990 to enlarge the Museum display space.

oversight led the Corps to join the Armed Forces Common Investment Fund (AFCIF), which has successfully managed the Corps' investments since that date.

THE ROYAL ENGINEERS OFFICERS' WIDOWS SOCIETY

An Officers' Widows Fund was originally established in 1783. The original basis was the payment of an annual subscription, based on rank, which provided an annual payment on

Above: Her Majesty the Queen, Colonel-in-Chief of the Corps, inspecting Captain Nick Beighton's canoeing gold medal from the Paralympics during her visit to the Corps in October 2016 to mark the 300th anniversary of its formation. The then Chief Royal Engineer, Lieutenant General Sir Mark Mans KCB CBE DL, greeted Her Majesty on arrival, remarking that the event commemorated three centuries of achievements that had seen Royal Engineers involved in virtually every scientific development and technical function of the Army, from mapping to fortifications, transport to communications, diving to flying, and from bomb disposal to high risk search. On other achievements, he highlighted the Corps' sporting successes from winning the FA Cup in 1875 to the recent individual performances by three Sappers at the Paralympic Games in Rio de Janeiro.

Below: The Royal Engineer Headquarter Mess* in Brompton Barracks, Chatham became such in 1856, when the Royal Sappers and Miners were amalgamated with the Royal Engineers. From 1807 until 1848 there had been a joint Mess for artillery and engineer officers. The dining room illustrated was designed by Captain Francis Fowke (see also page 97) and built in 1861. The north and south annexes behind the columns were later additions.

*This is the traditional designation, without the plural in either *Engineer* or *Headquarter*, and without the word *Officers*.

death that was the same for all ranks. The original pension was £30, which rose to £40 in 1812. Nowadays the Annuity Fund, to which members subscribe, has the aim 'to establish, maintain and administer a fund for the better support and maintenance of the spouses and dependants of Officers of the Corps of Royal Engineers'. The Samaritan Fund is for relieving 'poverty amongst widows and orphans of Officers of the Corps of Royal Engineers' and other legal charitable purposes.

THE INSTITUTION OF ROYAL ENGINEERS

The Institution's origins as a professional body began with the publication in 1837 of the first Professional Papers by Lieutenant William Denison (see page 94). By this means, ideas were shared between sappers around the World. In 1875, the Institute (as it was then known) was founded principally as a formal means of exchanging professional information and the Museum and Library, as a part of the Institution, have made a positive contribution to that. *The Royal Engineers Journal* replaced the *Professional Papers* in 1904. And, more recently, the RE Historical Society has been reformed to foster an interest in learning from the Corps' history. Another important part of the Institution's activities is the administration of awards and prizes for professional achievements by members of the Corps.

The Institution obtained its Royal Charter

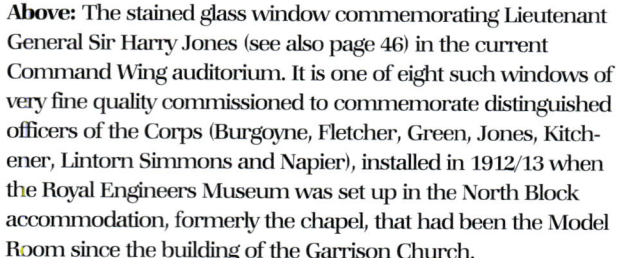

Left: *The Sapper* magazine was the brain-child of three corporals in 1895 who, with the help of Engineer Clerk Sergeant Hirst, prepared a manuscript for the first copy for approval by the Assistant Commandant at Chatham. The first print-run of 3,000 was sold out within a week. Thus began one of the most popular corps and regimental magazines that now runs to 6,500 copies every second month.

Above: The stained glass window commemorating Lieutenant General Sir Harry Jones (see also page 46) in the current Command Wing auditorium. It is one of eight such windows of very fine quality commissioned to commemorate distinguished officers of the Corps (Burgoyne, Fletcher, Green, Jones, Kitchener, Lintorn Simmons and Napier), installed in 1912/13 when the Royal Engineers Museum was set up in the North Block accommodation, formerly the chapel, that had been the Model Room since the building of the Garrison Church.

Above: Saint Barbara's Church dates from 1854 when the garrison outgrew the former chapel, which was inside the barracks. It was originally built as a combined chapel and school; a classroom during the week and a place of worship on Sundays. There was no music in the church until the Royal Sappers and Miners moved from Woolwich with their band. The church contains many memorials from the Indian Mutiny to the Falklands and other treasures such as some exceptional stained glass and paintings. It also incorporates a Naval Chapel and houses the Naval Rolls of Honour from both world wars.

in 1923. Its charitable objects are '… for the promotion and advancement of the science of military engineering and to promote military efficiency and particularly the military efficiency of the Corps of Royal Engineers'.

These imply a role in the professional development of its membership, until recently confined to the serving and retired commissioned officers of the Corps. As its membership expanded to include all ranks of the regular and reserve Corps, and with greater emphasis placed nationally on professional recognition, the Institution became a Licensed Member of the Engineering Council in 2007, joining the other Professional Engineering Institutions in being able to register its members as Engineering Technicians, and Incorporated and Chartered Engineers.

THE ROYAL ENGINEERS MUSEUM

The Royal Engineers Museum is acknowledged as one of the United Kingdom's leading military museums. Combining a library that was founded

Left: The V2 Rocket, one of a number recovered by Sappers at the end of WWII. (RE Museum gallery)

Below: The original model of Mulberry Harbour used for briefing the Prime Minister on the progress of the construction of the harbour after D-Day. (RE Museum gallery)

Above: An 1834 full dress coat of the Bengal Engineers.

Right: The Royal United Services Institution Gold Medal awarded to Captain W. Baker Brown for his military essay in 1898.

Left: The Canada General Service Medal 1866–70 was a medal issued some time after the event, in this case 29 years, little comfort to the many soldiers who would have died during the intervening period. The two clasps on this example are for the Fenian campaigns and the Red River expedition. The 'Fenian Brotherhood' was formed in the United States in 1858 and became the source of recruits for an army raised to attack British interests in Canada in 1866 and again in 1870. In that year also the inhabitants of the Red River, largely Métis, rebelled against the new province of Manitoba. The reverse of this medal shows the Canadian flag surrounded by a maple wreath with the word 'CANADA' above.

in 1813 and a museum that first opened to the public in 1912, it has preserved the heritage of the Corps and acted as a centre for the study of military science for over two hundred years.

Since the Royal Engineers was established in 1716, its members have recorded and collected evidence of their achievements. This developed into a Museum collection that grew alongside the RSME, being primarily established to support training, whilst 35 Corps Libraries supported Royal Engineer learning in depots across the globe. The museum and library now form a single unique collection, recording over three

hundred years of the nation's engineering and scientific evolution from a military perspective.

Of particular import is the world-class collection of Napoleonic War archives: blueprints and plans for the Mulberry Harbours and First World War Unit War Diaries as well as the hundreds of letters, diaries and drawings that record the individual Sapper's experience and expertise. Highlights of the object collection include Victoria Crosses marking the gallantry of twenty-five Royal Engineers, the Imperial Chinese costume of General Charles Gordon, the Brennan Torpedo and a war winning Bailey Bridge. The internation-

Below: The Churchill Bridgelayer is part of the outside display of the Royal Engineers Museum (2005).

Above right: The Waterloo Map, following restoration, in its new protective display, and the interactive computer based interpretation panel. The project was funded in part by the Friends of the RE Museum.

Right: A set of fortification drawings by Practitioner Engineer (later Lieutenant General) Charles Tarrant is one of the earliest sets of such drawings in the Museum collection. Illustrated is his plan of Dover Castle showing the batteries erected in 1756 and 1757. (See also page 12)

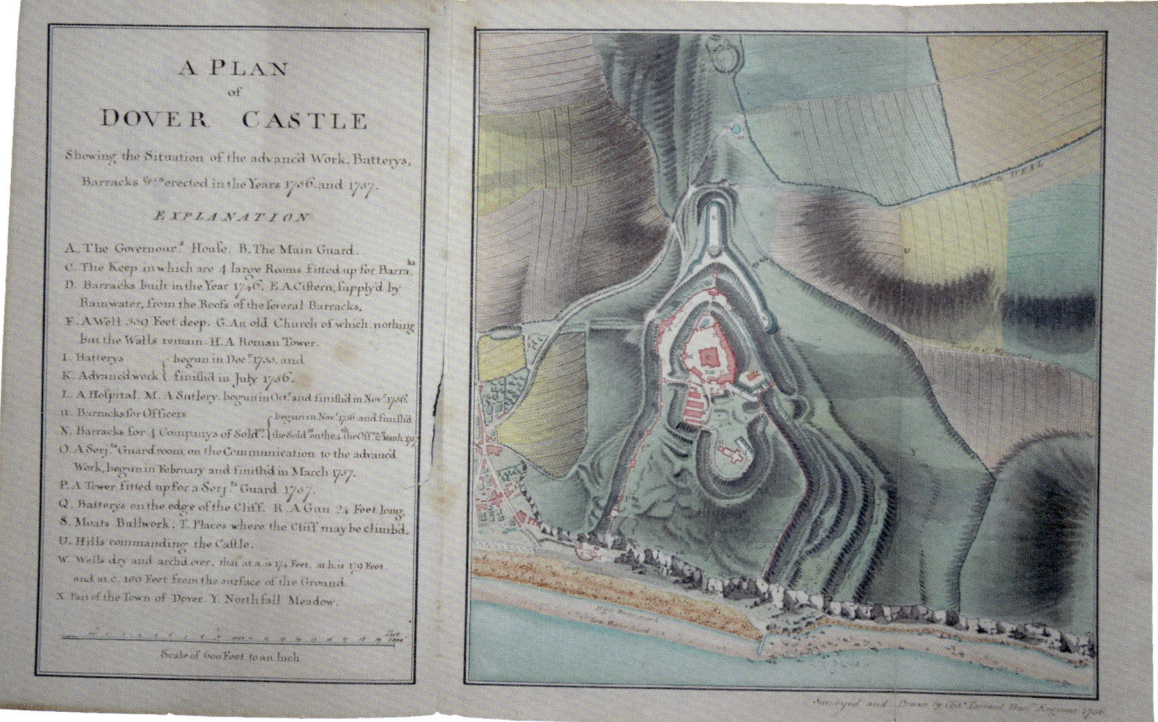

Right: The Royal Engineers Museum collection contains the work of many talented sappers who took their paints and brushes on duty overseas tours. This watercolour by Major General Sir John Ardagh is a particularly fine example.

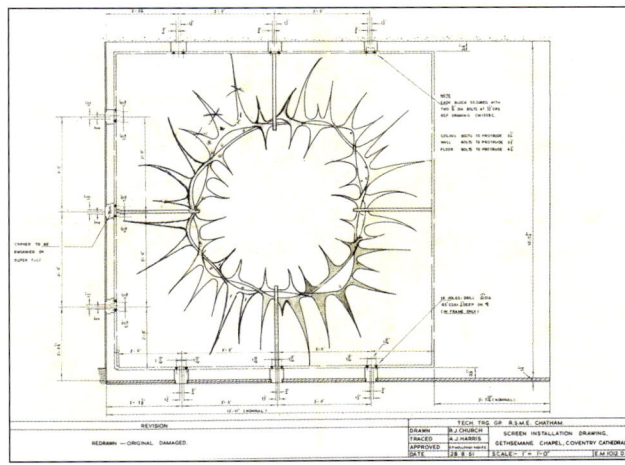

Above: The original drawing for the metal grille representing the Crown of Thorns, designed for the Gethsemane Chapel, Coventry Cathedral, by Sir Basil Spence RA. The grille was made in 1961 in the Blacksmiths' and Welders' shops of the Trade Training Workshops at Chatham. It was finished in the workshops of No. 1 Engineer Stores Depot, Long Marston. A sample 'thorn' approved by Sir Basil is also held in the Museum. The grille was a gift from the Corps.

Right: Volunteers studying the map and plan collection from the Royal Engineers Archive.

ally significant photography collection comprises an extensive archive of portrait photographs, extremely rare mid-19th century images from Africa, Tibet and Canada as well as significant 20th century collections including images from the 1945 liberation of the Bergen-Belsen concentration camp.

The Museum is constantly developing to meet the needs of a growing historical collection and changing public expectations. In 1985 it relocated from the old RSME chapel to the Ravelin Building, improving public access and creating space for extensive vehicle displays. The galleries were enlarged further when, in 1990, the building's original courtyard was covered and post-World War Two displays installed. Since then, a Model Bridge Centre and vehicle displays at the nearby Chatham Historic Dockyard have increased the displayed collection and the introduction of interactive and other new technologies have enhanced the Museum's galleries. Many of these project were made possible through the fundraising activity of the Friends of the Royal Engi-

Left: Flying Bridge Complete. Photograph from the Corps archive showing Sappers training on the Great Lines in Chatham, c1870. The archive holds tens of thousands of images of Royal Engineers at work and play, from the very early years of photography to the present day.

neers Museum and the, now disbanded, Royal Engineers Museum Foundation as well as the generosity of hundreds of visitors and donors.

In addition to its fascinating displays, the Museum now offers an extensive public service that includes vibrant and engaging family activi-

THE MEDAL COLLECTION
Throughout this book examples from the medal collection appear in the context of the
operations in which they were earned. Some of the more unusual ones are shown here.

Top: Showcase displaying the batons and honours of
Field Marshal Earl Kitchener (see page 92) in the Field
Mashall Room of the Museum (see text). (RE Museum
gallery)

Middle, left to right:
The Royal Guelphic Order was founded in 1815 by
HRH the Prince Regent (later King George IV). It took
its name from the family surname of the British sover-
eigns from George I onwards and was awarded by
the Crown of Hanover to both British and Hanoverian
subjects for distinguished service to Hanover. On the
death of William IV in 1837, it became a purely Hanoverian award. The neck badge of the military version is
that of Sir John Oldfield, one of the eleven Royal Engineers present at Waterloo.

The Edward Medal is awarded for acts of bravery in
factory accidents and disasters and was instituted in
1909. It was awarded in both silver and bronze. In
1971 living recipients were invited to exchange their
medals for the George Cross, but not all made the
exchange. The medal shown was awarded to Sergeant
Harris who was working at an anti-aircraft battery
near Faversham when there was an explosion and fire
at a nearby explosives factory. Harris took nine men
from the battery and went to help. He was awarded the
Meritorious Service Medal. However, because of the
example he had set to the civilians, who had been hesi-
tant at first, he was later awarded the Edward Medal as
well, the only sapper to have been so honoured.

The Polar Medal was instituted in 1904 by Edward VII
as an award for Arctic and Antarctic exploration. The
reverse depicts the ship *Discovery* with a sledging party
of six men in the foreground, and a heavily laden sledge
with a square sail. This medal, one of five received by
members of the Corps, was awarded to Warrant Officer
Class 2 I. Beney and has the clasp ANTARCTIC 1957–58.
It is the only Polar Medal in the collection.

Bottom: The Distinguished Service Cross, formerly
the Conspicuous Service Cross, instituted in 1901,
was intended for those warrant and subordinate
officers of the Royal Navy who were not eligible for
the Distinguished Service Order. In 1940 Army and
Royal Air Force officers and Warrant Officers serving
aboard Her Majesty's
ships also became
eligible for this award,
hence this award to
Warrant Officer Class 2
J H Phillips, for his
gallantry in bomb
disposal on board HMS
Argonaut (see also page
182). Sadly, this rare
medal is no longer in
the collection.

ties, diverse exhibitions and community engagement programmes, a schools' learning service specialising in science and engineering education, and support for University degree programmes and research facilities. The Museum staff and some dedicated volunteers continue researching and cataloguing the collection, improving online content, answering public enquiries and supporting preservation work. The collection continues to grow and the Museum works closely with the Corps to ensure that the experiences and achievements of contemporary Sappers will be remembered by future generations.

THE ROYAL ENGINEERS ASSOCIATION

The Royal Engineers Association dates back to the Royal Engineers Old Comrades Association, founded in 1912, whose name was changed to the Royal Engineers Association in 1952. The general aims of the Association were fostering *esprit de corps*, circulating information about the Corps to help potential recruits, helping to find employment for ex-members and bringing to the notice

of the Royal Engineers Charitable Fund any deserving of assistance. That fund was founded in 1868 and re-designated the Royal Engineers Benevolent Fund in July 1943.

From that arose the current flourishing Association whose primary focus is the provision of welfare and benevolence assistance to the Sapper Family across those serving, veterans and dependants; wherever there is a need the Association is there to provide support.

Annually, the Association provides around £1 million in benevolence and welfare support for up to 600 individuals who find themselves in need of direct financial support and for the numerous units, branches and individuals who benefit from the Associations funding of welfare facilities, including communal recreation areas,

Below: Support to Esprit de Corps. The REA engages with serving soldiers to support *esprit de corps* and publicising the work of the Association.

Left: The REA Badge of Merit in recognition of valuable service.

Above: The Association Standard Bearers at the Sapper Memorial in the National Memorial Arboretum.

Corporals clubs, adventure training and sport. The Association brings employment opportunities to service leavers through Sappers Network.

The vibrant branch structure has more than 100 branches nationally and internationally, providing a vital source of friendship and support, linking the serving corps with veterans and families. With different types branches, from local to national groups, such as Airborne and Commando, to special interest, such as football and cycling, there are branches to suit all needs. There are numerous national and local

Left: First Light was commissioned by the Royal Engineer Yacht Club from the artist F. H. Wagner to commemorate the 150th anniversary of the founding of the club in 1846. It shows the club yacht *Ilex* rounding the Fastnet Rock in 1926, the year that she won the Fastnet race. This was the year that *Ilex*, a twenty-ton yawl, was acquired by the Club and had a long and distinguished career, only being finally sold in 1950.

Left: The Royal Engineers Football team of 1871/72. Eight members of this team were also members of the 1875 team that won the Football Association cup.

events including annual weekends in Gibraltar and Chatham that see the Sapper family gathered together to celebrate in a spirit of comradeship and service before self.

SPORT AND ADVENTUROUS TRAINING

The Royal Engineers Games Fund was founded in 1879 primarily to help finance entertainment for Corps matches at Chatham. Officers from outside Chatham contributed to that end. Under the same principle a proportion of officers' and soldiers' subscriptions to Corps Funds is allocated to sport, games and adventurous training.

Sports that relate to the professional work of the Corps such as sailing, canoeing and mountaineering, and more recently parachuting and hang-gliding, have figured strongly in Corps activities.

The Royal Engineer Yacht Club dates back to 1846, by which time several yachts were plying the Medway and beyond in the hands of officers of the Corps. The Yacht Club also embraced rowing until 1950, when the separate Royal Engineer Rowing Club was established. The Yacht Club, which has the rare distinction of being permitted to wear the blue ensign undefaced, has provided generations of sappers of all ranks with exciting and affordable

Above: A replica of the Football Association Challenge Cup won by the Royal Engineers in 1875. It was presented to the Royal Engineers Football Club in 1972 to mark the centenary of the event.

Above: The Junior Leaders Centrepiece captures the spirit of adventure training in the Corps, with a climbing scene and panels depicting other activities. The Junior Leaders Regiment existed from 1958 to 1991 and fulfilled its role of providing the Corps with a whole generation of well-motivated young men.

Above: Two examples of adventurous training in the Corps.

Left: The Rudge Whitworth Cup is one of the oldest sporting trophies held by the Corps. It was presented by the Rudge Whitworth company in 1904 to the Royal Engineers Ubique Cycling Club.

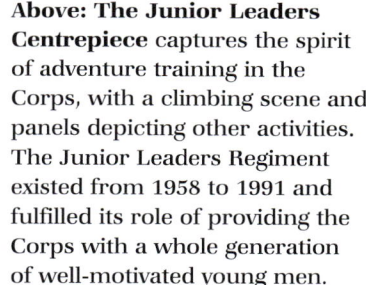

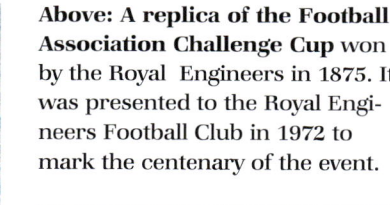

sailing both offshore and in dinghies. Many are the triumphs achieved in racing, both in club boats and those of private members, not least the 1926 victory in the Fastnet race (see illustration).

The Corps continues to support 'sport for all' and this is reflected in the wide ranging success enjoyed by a wide variety of teams and individuals. The Corps football, rugby (union and league) and cricket teams have regularly occupied the top spot in inter-Services and Army competitions and Corps boxing flourishes for both men and women. There has been a steady growth in women's sports and the recent success of the women's triathlon and netball teams has been particularly evident. Corps support to individuals has seen significant success for both the regulars and reserves in martial arts and the new sport of power lifting.

On the adventurous training front, the Corps remains at the forefront of mountaineering and has supported major expeditions to South America, the Himalayas and Greenland. Members of the Corps have rowed the Atlantic and trekked across the globe. There has been a steady trickle of support for extreme sports such as the 5-day Marathon des Sable in Morocco. More recently, the Corps has featured strongly in the Invictus* Games and Paralympics.

THE ROYAL ENGINEERS BAND

The Band can trace its origins back to 1850 when it was established as a brass band at Woolwich at the Depot of the Royal Sappers and Miners. When the Depot moved to Chatham in 1856, so did the Band (see illustration). Only in 1948 did the Band become an official unit of the Army. The following year a second Band was approved for Aldershot. These Bands were amalgamated

*The word invictus means unconquered and the Games harness the power of sport to inspire recovery, support rehabilitation and generate understanding and respect for those who serve their country.

in 1985. In 1994, the establishment of the Band was reduced from forty-nine to thirty-five and all members of the Band were transferred to the Corps of Army Music, including the responsibility for the funding of uniforms and instruments.

The Band travelled extensively around the world and performed in recent years in the Falkland Islands, Australia, Hong Kong, the Middle East and Europe. It was the first British military band to perform in the former Soviet Central Republic of Uzbekistan and, in 2000, it performed in South Korea as part of an international Tattoo to commemorate the fiftieth anniversary of the start of the Korean War. The Band has also played at the opening of both the Channel Tunnel and the Queen Elizabeth II Bridge, Dartford.

The Band was mobilized in its wartime medical role and deployed to Iraq in 2003.

In 2019, the Royal Engineers Band was amalgamated with the Royal Artillery and Adjutant General's Bands to form the British Army Bands, Tidworth. The tradition of over 150 years of the Corps Band is now maintained by Volunteers in the Nottinghamshire Band of the Corps of Royal Engineers.

Left: The Royal Engineers Band seen on the Brompton Barracks square in *c*.1878.

Left: The Royal Engineers Band in 2016 at the end of the Parade to mark the Queen's visit to the Corps in its 300th year since formation.

Far Left: Members of the Royal Engineers Band in their war role in Iraq. The photograph was taken in early March 2003 during training.

THE CORPS BADGE (THE ROYAL ARMS)

On 10 July 1832 King William IV granted the Royal Regiment of Artillery and the Corps of Royal Engineers permission to wear on their appointments the Royal Arms and Supporters, together with a cannon and the motto 'Ubique quo fas et gloria ducunt'.* In 1868 the cannon was omitted from the Corps Badge. The Corps Badge used to be worn on an officer's sabretache and cartouche, and on parts of his charger's saddlery. It was worn by all ranks of the Corps of the The Regular Army on their full-dress blue spiked helmet until 1914. The Royal Engineers Militia, Volunteers and Territorials however, had a slightly different badge in that the motto Ubique was omitted from the scroll under the Royal Arms and its place taken by a laurel branch.

*'Everywhere where Right and Glory lead'. By custom and practice this has been separated into two mottoes.

THE CAP BADGE

In 1782 the device worn on officers' sword belts was the King's cypher with the crown over it. At some time later the cypher was surrounded by the garter, on which was placed 'Corps of Royal Engineers'. This was later changed again to simply 'Royal Engineers'. A similar device was worn on the breastplate of cross belts introduced for the Royal Sappers and Miners in 1823. It is not known when the laurel wreaths were added but it was probably in recognition of the work clone by the Corps during the Napoleonic Wars. The Cap Badge was first used as a hat badge on the khaki helmets issued to troops during the South African War of 1899 - 1902. *Description:* The Garter and Motto surmounted by a Crown; within the Garter the Royal Cypher; without the Garter a wreath of laurel; on a scroll at the bottom of the wreath, Royal Engineers. The Garter, Motto, Royal Cypher, crown and scroll are in gilt, and are raised above the laurel wreath. The laurel wreath is in silver plate. The soldiers' badge is similar but in gilding and white metal.

THE MONOGRAM OR CYPHER

The Monogram, or Cypher. Is used on notepaper, Christmas Cards and other similar documents. It is not worn on uniform but is emblazoned on the Fanfare Trumpet Banners of the Royal Engineers Band.

ROYAL ENGINEER GRENADE

An embroidered grenade was first worn on the tail of an RE Officer's full dress scarlet coatee in 1824. The following year a brass grenade was introduced for other ranks of the Royal Sappers and Miners. The grenade was later worn on the epaulet and then on the collar. The number of flames to the grenade has varied, but in 1922 a nine-flamed grenade, with the motto 'Ubique' below it, was authorized. The Royal Artillery grenade is similar, but has only seven flames.

WINGS

The exact origin of 'Wings' as the Corps Regimental March is obscure, but in 1870 the Commandant of the School of Military Engineering, unaware that the Corps had been allocated 'The British Grenadiers' by the War Office, directed that the Band Committee should adopt a popular air of the day as the Regimental Quick March. The Committee adopted Wings. It is a combination of two tunes, scored by Bandmaster Newstead of the Royal Engineers Band, one being from the air 'The Path Across the Hills', a tune of unknown German origin, and the other 'Wings', a contemporary popular song by Miss Dickson. It was not until 1902 that Wings was also officially recognised.

Wings to bear me over mountain and vale away;
Wings to bathe my spirit in morning's sunny ray;
Wings that I may hover at morn above the sea;
Wings through life to bear me, and death triumphantly.

Wings like youth's fleet moments which swiftly o'er me passed;
Wings like my early visions, too bright, too fair to last;
Wings that I might recall them, the loved, the lost, the dead;
Wings that I might fly after the past, long vanished.

Wings to lift me upwards, soaring with Eagle flight;
Wings to waft me heav'nwards to bask in realms of light;
Wings to be no more wearied, lulled in eternal rest;
Wings to be sweetly folded where Faith and Love are blessed.

HURRAH FOR THE CRE!

The Corps Song originated among sapper units during the South African War. The words, partly in English and partly in Zulu, are sung to the tune of the traditional South African song *Daer de die ding*. The Zulu words are a complaint that as there is too much work for too low wages and little food, and they are leaving.

Ooshta	South African native working cry
Ikona malee	No money
Picaninny skoff	Little food
Ma-ninga sabenza	Lots of work
Oolum-da	South African native working cry

Good Morning Mr Stevens and windy Notchy Knight,
  Hurrah for the CRE
We're working very hard down at Upnor Hard,
  Hurrah for the CRE
You make fast, I make fast, make fast the dinghy,
Make fast the dinghy, make fast the dinghy,
You make fast, I make fast, make fast the dinghy,
Make fast the dinghy pontoon.
For we're marching on to Laffan's Plain,
To Laffan's Plain, to Laffan's Plain,

Yes we're marching on to Laffan's Plain
Where they don't know mud from clay.
Ah, ah, ah, ah, ah, ah, ah, ah,
Ooshta, ooshta, ooshta, ooshta,
Ikona malee, picaninny skoff,
Ma-ninga sabenza, here's another off.
Oolum-da cried Matabele,
Oolum-da, away we go.
Ah, ah, ah, ah, ah, ah, ah,-
Shush .... Hurrah!

FURTHER READING

CORPS HISTORY IN GENERAL

The History of the Corps of Royal Engineers
(Volumes 1 to 12, 13 in preparation)

Up to 1914. Volumes 1 and 2, published in 1889 and written by Major General Whitworth Porter, cover events from the Norman Conquest to the 1882 war in Egypt. Volume 2 includes Corps organisation and the civil work of the Corps for the same period. Volume 3, published in 1915 and written by Colonel Sir Charles Watson, who died the same year, takes the story up to the end of the South African War. Volume 4, written by Brigadier General W. Baker Brown, was not published until 1952. It fills a number of gaps that Volume 3 should have included but for Colonel Watson's death and takes the narrative to the beginning of the First World War.

First World War and the inter-war years. Volumes 4, 6 and 7, contributed to by numerous officers and edited by Major General H. L. Pritchard, covers the whole of the First World War and the inter-war period to 1938; it was also not published until 1952.

Second World War. Volumes 8 and 9 were written by Major General R. P. Pakenham-Walsh and published in 1958. They cover the whole war and the immediate aftermath, to 1948.

Post-world war years. In the post-war years Volumes 10 (*1948 to 1960*, published 1986), 11 (*1960 to 1980*, published 1993) and 12 (*1980 to 2000*, published in 2011) were contributed to by a number of authors under the chief editorship of Colonels Ian Wilson and Hugh Mackintosh and Brigadier Alasdair Wlson, respectively. Volume 13 is in preparation under the general editorship of Major General Mungo Melvin.

Indian Engineers. These volumes of Corps History do not include the work of the Indian Engineers except where they were directly involved with the Royal Engineers. Their distinguished history is covered in Lieutenant Colonel E. W. C. Sandes' books published by the Institution (*see Select Bibliography*).

Short Histories

Derek Boyd's *The Royal Engineers*, published by Leo Cooper in 1975 in the 'Famous Regiments' series, is a good summary of the achievements of the Corps but is now out of print. Major Peter Aston's *The Royal Engineers, a Short*

History, originally published in 1993 by the Institution and regularly updated, remains the official version of its kind.

JOURNALS AND UNPUBLISHED WORKS

This book has drawn extensively on *The Royal Engineers Journal* and to a lesser extent on the *Professional Papers of the Royal Engineers* and *Sapper* magazine, available in the Royal Engineers Library. Lieutenant Colonel P. H. Kealy's *Follow the Sapper*, with its comprehensive research of the early 19th century Corps, is a valuable source. The author died before it was published, and it is hoped that this publication honours Colonel Kealy's work by taking the name he used. Other unpublished references of particular value in the Royal Engineers Library include:

Military Engineering in the Peninsular War. A digest of references by Major J. T. Hancock

Royal Engineer Architects. A PhD thesis by John Weiler covering the contribution of the Corps to building construction in the 19th century.

Roll of Officers of the Royal Engineers by Captain T. W. J. Conolly.

Portraits, Pictures and Silver of the Royal Engineer Head-quarter Mess, Chatham, produced by the Institution of Royal Engineers in a limited edition.

SELECT BIBLIOGRAPHY

Barton, Peter, and Johan Vandewalle, *Beneath Flanders Fields*, Spellmount 2004.

Beanse, Alec, *The Brennan Torpedo*, The Palmerston Forts Society, 1997.

Berthon and Robinson, *The Shape of the World*, George Philip, 1991.

Bowen, Brigadier D. H., *Queen's Gurkha Sapper*, Institution of Royal Engineers, 1998.

Bradbury, Jim, *The Medieval Siege*, The Boydell Press, 1998.

Brazier, Lieutenant Colonel C. C. H., *XD Operations*, Pen & Sword, 2004.

Broke-Smith, Brigadier P. W. L., *The History of Early British Aeronautics*, a reprint of a series of articles from the *Royal Engineers Journal*.

Cima, Keith H., *Reflections from the Bridge*, Baron Birch, 1994.

Conolly, Quartermaster-Sergeant T. W. J., *The History of the Royal Sappers and Miners*, Longman, 1855.

Cooper, General Sir George and Major David Alexander (eds.), *The Bengal Sappers*, Institution of Royal Engineers, 2003.

Foulkes, Major General C. H., *'Gas!' – The Story of the Special Brigade*, Blackwood, 1934.

Grieve, Captain W. Grant and Bernard Newman, *Tunnellers*, Herbert Jenkins Ltd, 1936.

Hill, Beth, *Sappers: The Royal Engineers in British Columbia*, Horsdal & Schubart, 1987.

Hodson, Yolande and Alan Gordon, *An Illustrated History of Military Survey*, Military Survey Defence Agency, 1997.

Jackson, General Sir William, *The Rock of the Gibraltarians*, Associated University Presses, 1986.

Joiner, Colonel J. H., *One More River to Cross*, Pen & Sword, 2001.

Jones, Major General Sir John T., *Journal of the Sieges carried on by the Army under the Duke of Wellington, in Spain, during the years 1811 to 1814*, London, 1846.

— *Memoranda Relative to the Lines thrown up to cover Lisbon in 1810*, London, 1846.

Kealy, Lieutenant Colonel P. H., *Sir Charles Pasley and the School of Military Engineering*, published posthumously by the Institution of Royal Engineers.

Kightly and Chèze Brown, *Strongholds of the Realm*, Thames & Hudson, 1973.

Legget, Robert, *John By, Builder of the Rideau Canal, Founder of Ottawa*, Historical Society of Canada, 1982.

Lyall Grant, Major General Ian, *Burma: The Turning Point*, Zampi Press, 1993.

Nalder, Major General R. F. H., *The Royal Corps of Signals. A History of its Antecedents (circa 1800–1955)*, Royal Signals Institution, 1958.

Napier, Gerald, *The Sapper VCs*, The Stationery Office, 1998.

Nowers, Colonel John, *Steam Traction in the Royal Engineers*, North Kent Books, 1994.

Regimental Headquarters Royal Engineers, *Corps Memoranda*.

Royal Bombay Sappers and Miners Association, *The Royal Bombay Sappers and Miners 1939–1947*.

Sandes, Lieutenant Colonel E. W. C., *The Military Engineer in India* (two vols), Institution of Royal Engineers, 1935.

— *The Indian Sappers and Miners*, Institution of Royal Engineers, 1948.

— *The Royal Engineers in Egypt and the Sudan*, Institution of Royal Engineers, 1937.

Sinclair, Major General G. B., *The Staff Corps*, Institution of Royal Engineers, 2001.

Smithers, A. J., *Honourable Conquests*, Leo Cooper, 1991.

— *A New Excalibur*, Leo Cooper, 1986.

Ward, *The School of Military Engineering, 1812–1909*.

Wells, Edward, *Mailshot: A History of the Forces Postal Service*, Defence Courier and Postal Services Royal Engineers.

ACKNOWLEDGEMENTS

Many people contributed to the first edition, and the Institution is particularly grateful to the late General Sir George Cooper for his constructive comments and to those who contributed material not otherwise acknowledged, particularly Captain Graham Hornby (medals and main scripts), Colonel Jim Joiner (bridging), Captain Andrew Phillipson (Revolutionary Wars), Major Alan Gordon and Dr Yolande Hodson (Survey), Captain Simon Fenwick (Postal and Courier), Colonel John Nowers (steam traction), Colonel Tony George and Dr Eric Old (Royal Monmouthshire Royal Engineers), Major Arthur Hogben (Bomb Disposal) and Brigadier Bill Woodburn (Chitral).

In producing this second edition, the Institution is indebted to Colonel David Hindle, of the Engineer and Logistic Staff Corps, whose generous donation made this project possible. Major General Mungo Melvin gave valuable advice on the additional material in this volume and kindly volunteered to write the foreword. Charles Holman, the Corps Secretary, Ken Kirk, the Controller of the REA, and Rebecca Nash, the Director of the RE Museum, each contributed to the revision of Chapter 49 to ensure their areas were accurately represented.

INDEX

SURVEY

1747 Military Survey started

For five-and-a-half centuries since the Norman Conquest engineers were on the pay-roll of the sovereign first as 'kings' engineers' and later, after the introduction of gunpowder, as engineer officers of the Board of Ordnance. 'Ordnance Trains' of artillerymen and engineers would be assembled to go to war with the Army (see Chapters 1 and 2). In 1716 the Board of Ordnance gave birth to the Royal Regiment of Artillery leaving the Corps of Engineers as a separate entity.

General David Watson

Manhattan c.1770

Palestine survey

INDIAN ENGINEERS

To Royal Engineers

Royal Military Canal

ROYAL STAFF CORPS

1799 Royal Staff Corps established

c.1830 Functions to Royal Engineers

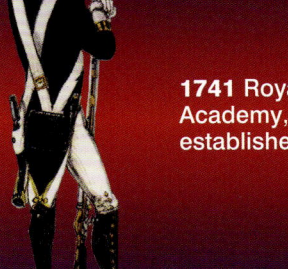

1716 Royal Artillery and Corps of Engineers placed on separate establishments

1757 Military rank for Corps of Engineers

1741 Royal Military Academy, Woolwich, established

1787 Royal Engineers given 'Royal' title

1787 Royal Military Artificers formed

1797 Soldier Artificers and Royal Military Artificers amalgamated

1812 Royal Engineer Establishment founded at Chatham

1813 Royal Sappers and Miners formed

1827 Siege of Bhurtpore

1809 Lines of Torres Vedras

1813 Crossing of the Adour

1830 Royal Sappers and Miners in Canada

1855 First use of tele-graph

CORPS OF

1704 Battle of Blenheim

1711 Lines of Non Plus Ultra

1702–13 War of the Spanish Succession

1756–63 Seven Years War

1772

SOLDIER-ARTIFICERS, ROYAL MILITARY GIBRALTAR ARTIFICERS

1775–83 War of American Independence

1787

ROYAL MILITARY ARTIFICERS

1797

1803–15 Napoleonic Wars

1803–6 Second Maratha War

1813

1812–15 War of 1812

ROYAL SAPPERS AND MINERS

1839–42 First Afghan War

1845–52 Sikh Wars

1855 Royal Sappers and Miners amalgamate with Royal Engineers

1854–6 Crimean War

1857– Indian Mutin[y]

1066–82 Norman conquest	
1078 Bishop Gundulf's White Tower	
1180 Dover Castle inner keep	
1337–1453 Hundred Years War	
1346 Battle of Crécy, first use of cannon	
1414 Office of Ordnance	
1513 Battle of Flodden	
1642–51 English Civil Wars	
1698 Ordnance Trains raised	

1779–83 Siege of Gibraltar

1793–1801 French Revolutionary Wars

Torres Vedras

The Duke of Marlborough and Colonel John Armstrong

*Lying in wa[it]
Lucknow 1[...]*